SpringerBriefs in Physics

For further volumes:
http://www.springer.com/series/8902

Marcos d'Ávila Nunes

Hadron Therapy Physics and Simulations

 Springer

Marcos d'Ávila Nunes
University of São Paulo
Ribeirão Preto
Brazil

ISSN 2191-5423   ISSN 2191-5431 (electronic)
ISBN 978-1-4614-8898-9   ISBN 978-1-4614-8899-6 (eBook)
DOI 10.1007/978-1-4614-8899-6
Springer New York Heidelberg Dordrecht London

Library of Congress Control Number: 2013947925

Printed on acid-free paper

Springer is part of Springer Science+Business Media (www.springer.com)

*I am deeply grateful to my wife Maria Silvia
and to my daughters, Alessandra, Ariane,
Carla, Rafaela, and Samantha, for taking
care of me with extreme dedication while my
health was critical because of bowel cancer,
and for saving my life and always
encouraging my work.*

*To Professor José Antônio Mansur Mendes,
worldwide exponent of gastroenterology,
I offer a tribute and thanks for extending me
his hands, as does my family; "Por você eu
faço!" and for releasing me from Dante's hell.*

*"I cross my arms on the table, put my head
on my arms,
I must want to cry, but I cannot get the tears…
As much as I strive to have great pity for me,
I do not cry,
My soul cracked under the curved indicator
that touches…
What's become of me? What's become of me?"*

(Fernando Pessoa)

*My interest is to help those who suffer,
providing useful information such as that
regarding hadron therapy, for alleviating
their suffering, their pain, and making them
happy again.
I've been through this and I was helped,
I was saved, now I wish to help others
with a similar fate…
Blessed are those who prepare thoroughly
to help others.*

(Marcos d'Ávila Nunes)

Preface

This Springer Brief fills a gap in the medical literature concerning the treatment of tumors using hadron therapy in South America, with a particular focus on Brazil. Hadron therapy uses a particle beam to treat tumors located close to critical structures in the body, and tumors that respond poorly to conventional radiotherapy. One of the reasons for writing this brief came from the need to collaborate with researchers, physicians, biomedical physicists, engineers, radiobiologists, undergraduate students and graduates in the field of high energy physics, and other individuals interested in hadron therapy. Carbon ion beams are effective in the control of radio-resistant tumors (10 % of all solid tumors) such as those found in the brain, lung, and liver. These tumors respond well when bombarded with these ions. Each carbon nucleus deposits 24 times more energy in a cell than a proton, and has a higher radiobiological effectiveness (RBE). The reason for this higher RBE is the high ionization density produced by the carbon ion as it traverses a cell, and this results in more disruptive damage to the DNA double helix. The oxygen enhancement ratio (OER) refers to the enhancement of the therapeutic effect of ionizing radiation due to the presence of oxygen. Additional oxygen abundance creates additional free radicals and increases the damage to the target tissue. In solid tumors, the inner regions can be less well oxygenated than normal tissues, and a dose of radiation up to three times higher is required to achieve the same tumor control probability as is the case in tumors with normal oxygenation. The dependence of RBE and OER on linear energy transfer (LET) was studied by Brendsen in the early 1960s. He showed that the RBE reached a maximum at a LET of 100–200 keV/μm, the same LET at which the OER had decreased to approximately 1.0. More recent studies showed that the RBE "peaks" at a LET that is particle dependent, indicating that LET alone does not adequately define the microscopic energy deposition and its influence on biological effect. One of the complications associated with heavy ion and pion beams is the increase in RBE with depth in the stopping region. With heavier ions, the RBE is also dependent on dose and on the dose fractionation scheme used. The lack of information regarding hadron therapy has led to unsuitable recommendations and retreatment with conventional radiotherapy in cases of cancer recurrence. This has prevented adequate treatment using hadron therapy with carbon ions, resulting in patients having to undergo chemotherapy as a single therapy, which is not always satisfactory. In general, proton therapy centers in the USA do not accept patients who have undergone two (or more) courses of conventional radiotherapy; however, as protocol standards vary between proton therapy centers, desperate patients who have been refused treatment search for a different center in the hope that it will accept them. We have observed this type of behavior in Brazil because we do not have hadron therapy facilities. After receiving two or more doses of conventional radiotherapy, some patients come to us seeking help in finding additional therapy. It is regrettable that hadron therapy is currently not available in Brazil. Many have struggled to achieve the deployment of at least one cyclinac accelerator for the treatment of cancer patients, and the establishment of research and development in the field of hadron therapy. New emerging technologies such as the dielectric wall accelerator (DWA) are still in the development phase and have not reached the specifications required for clinical application. With the publication of this brief, we hope to establish collaborations with other centers to enable the definitive implementation of hadron therapy in Brazil.

I would like to express my sincerest gratitude to Dr. Harry Blom for his support, suggestions, and constructive criticism throughout this work. When I approached Springer-Verlag with a proposal for the publication of my book entitled *Large Hadron Collider—New Era of Discoveries*, Dr. Blom's attention was drawn to the chapter on hadron therapy. He suggested that I examine this issue more thoroughly and submit a new proposal. In line with his suggestion, I proceeded to work tirelessly on this subject for 18 months. It brought me great joy because I believe that this Brief will help many people in need of hadron therapy in Brazil. In addition, international contacts intensified rapidly during the course of writing this manuscript. I would also like to thank Jennifer Satten, who is an Associate Editor at Springer, for her invaluable help in the formation of the Springer Brief. Two masters students in physics showed great interest in the Brief; Lucas Burigo is from the Universidade Federal do Rio Grande do Sul (UFRGS, Brazil) and is currently working at the Frankfurt Institute for Advanced Studies (FIAS, Germany), and Thiago Miranda Viana Lima is working at the European Organization for Nuclear Research (CERN, Switzerland). Both students are in the process of completing PhD theses. I offer them my special thanks for the information that they provided and also for the encouragement that they gave me. I must also thank Christine Romero, MPH, RN, and Gayle Wuchenich, RN, both from the International Patient Services Department at Loma Linda University Medical Center, CA, USA, for detailed and extremely useful information regarding proton therapy centers within and outside of the USA, and for additional recommendations, videos, and references. I would also like to extend my thanks to Professor Don Lincoln for carefully answering all of my emails and for his collaboration, which has helped to dispel my doubts. Professor Lincoln is a great experimental physicist and was one of those responsible for the discovery of the top quark at FERMILAB. He is best known for his work in high energy physics, as a great educator and as an international speaker, worthy of belonging to the Nobel Club. I would also like to thank Professor Ugo Amaldi, President of the TERA Foundation, for permission to use figures from their research.

Reading his biography, it is easy to see the idealist who was always making the right choices and carrying out original research in the field of hadron therapy. He is an excellent international speaker and a motivator of young researchers; he has also saved lives with the establishment of the Centro Nazionale di Adroterapia Oncologica (CNAO) in Italy. Thanks also go to Dr. Andrea Mairani for permission to use the diagram relating to the FLUKA particle transport code, for facilitating an understanding of the use of this simulation code, and also for supporting the establishment of new hadron therapy centers worldwide. I would like to thank Dr. Annette Tuffs, Head of Corporate Communications/Press Office, for permission to use the figures regarding the Heidelberg Ion-Beam Therapy Center (HIT), together with a detailed explanation of its components (source: Heidelberg University Hospital). Many thanks to Dr. George J. Caporaso, inventor of the dielectric wall accelerator (DWA) and holder of numerous patents, for permission to use the figures showing the DWA at a fairly advanced stage of construction. We will soon have a more affordable linear accelerator, which will help in saving thousands of lives. Professor Yoshinori Fujiyoshi kindly allowed me to use the figure regarding aquaporins and also offered encouragement. Dr. Marco Antonio Hakime gave his support and encouragement while I was hospitalized, and showed a great interest in hadron therapy. Finally, thanks go to Miss Ana Luiza Menezes Baldin for her translation work, and for understanding that we need advanced physics techniques to fight cancer.

Ribeirão Preto, Brazil   Marcos d'Ávila Nunes

About the Author

Marcos d'Ávila Nunes is an Associate Professor at the University of São Paulo (USP), SP, Brazil. He graduated from the Faculty of Medicine of Ribeirão Preto/ USP in 1967. He undertook postgraduate studies in advanced theoretical physics at the School of Engineering, São Carlos/USP (1968–1969). He received the title of Doctor of Science and Associate Professor from the Institute of Biomedical Sciences/USP in 1972 and 1977, respectively. Professor d'Ávila Nunes worked on his postdoctoral thesis for 5 years at the Massachusetts Institute of Technology (MIT, 1974–1975), Boston University Medical Center (BUMC, 1974–1975), National Institutes of Health (NIH, 1976), American National Red Cross (ANRC, 1976), University of South Carolina (USC, 1981–1982), and Wayne State University (WSU, 1982). He is known worldwide for his electrophysiological studies regarding biological membranes, the application of the thermodynamics of irreversible processes in membranes and also for the development of electron paramagnetic resonance (EPR) for the study of the behavior of molecular components of the biological membrane, subjected to a drop/rise in temperature and for simulations (molecular cryobiology—spin labeling). Professor d'Ávila Nunes has also published studies on medical informatics. He has participated in national and international congresses in the position of Chairman of Molecular Biophysics. Professor d'Ávila Nunes was responsible for seven courses involving postgraduate studies at USP. Having now retired from USP as an Associate Professor, he has decided to devote himself to writing books on nuclear medicine, including hadron therapy, advanced physics, large hadron collider, and a literary deal. These take advantage of the data relegated to his memory and his national and international experience in the field of basic science.

Contents

Chapter 1
Introduction

1.1 Brief History of Radiotherapy

Approximately 117 years ago Wilhelm Konrad Röntgen [9] discovered X-rays. He gave them this name without taking into consideration their nature. A few months after Röntgen's discovery, researchers concluded that X-rays could be used for diagnostic and therapeutic purposes. This discovery profoundly changed the medical applications of physics, and in 1901 earned him the award of the first Nobel Prize in Physics (Fig. 1.1).

Fig. 1.1 Wilhelm Konrad Röntgen, (1845–1923). Wikimedia Commons

M. d'Ávila Nunes, *Hadron Therapy Physics and Simulations*, SpringerBriefs in Physics, DOI 10.1007/978-1-4614-8899-6_1, © Marcos d'Ávila Nunes 2014

Shortly after this discovery, Henri Becquerel [10] discovered natural radioactivity. At his suggestion, this was studied by Madame Marie Curie and her husband (Fig. 1.2). Madame Curie [11] won two Nobel Prizes, one in Physics with her husband (1903) and another in Chemistry (1911). In addition, one of her two daughters, Irène, won a Nobel Prize in Physics; they were a real Nobel family.

Fig. 1.2 Marie Curie (1867–1934). Wikimedia Commons

In 1913, William D. Coolidge [7] carried out the groundwork for currently used X-ray techniques, developing a vacuum tube containing a tungsten hot cathode. Artificial radioactivity was discovered by Irene Curie [8] and her husband, and by Enrico Fermi [13] and collaborators (Fig. 1.3).

Fig. 1.3 (**a**) Irène Curie (1897–1956) with Jean Frédéric Joliot (1900–1958). (**b**) Enrico Fermi (1901–1954). Wikimedia Commons

In 1930, Ernest Lawrence [14] invented the cyclotron, a circular accelerator in which the particles have a spiral trajectory. This allowed for the creation of isotopes (Fig. 1.4).

Fig. 1.4 Ernest Lawrence (1901–1958). Wikimedia Commons

Nuclear reactors were developed to obtain isotopes intended for use in the medical field (Fig. 1.5).

Fig. 1.5 Core of a small nuclear reactor used in research. Wikimedia Commons

In 1950, positron emission tomography (PET) was created conceptually. This modality allowed for the scanning of the entire body to locate tumors, producing a three-dimensional image when combined with computed tomography (CT). Tumor tissue showed up in black on the image. In Fig. 1.6, the progressive disappearance of the tumor can be seen after treatment with chemotherapy for 4 months.

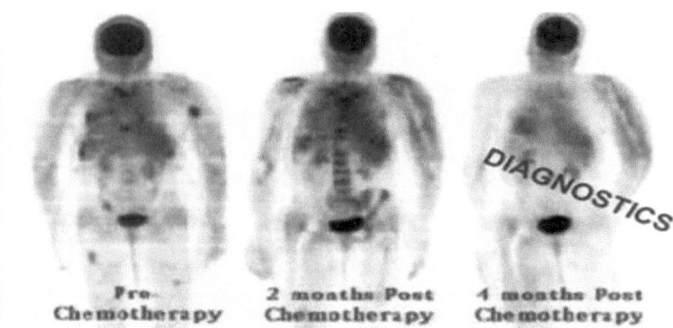

Fig. 1.6 Images from a whole-body PET scan. From *left* to *right*: pre-chemotherapy, 2 months after the start of chemotherapy and 4 months after the start of chemotherapy. Courtesy Ugo Amaldi [2]

The following images (Figs 1.7 and 1.8) are of PET and PET/CT devices.

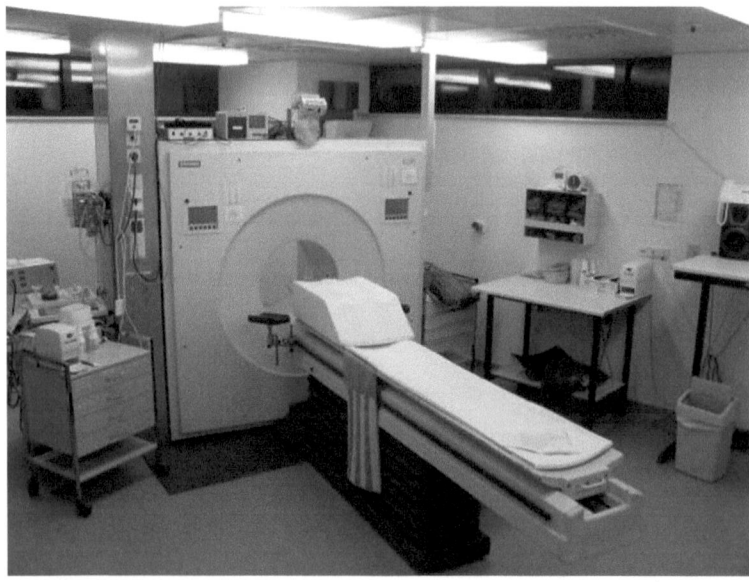

Fig. 1.7 PET device

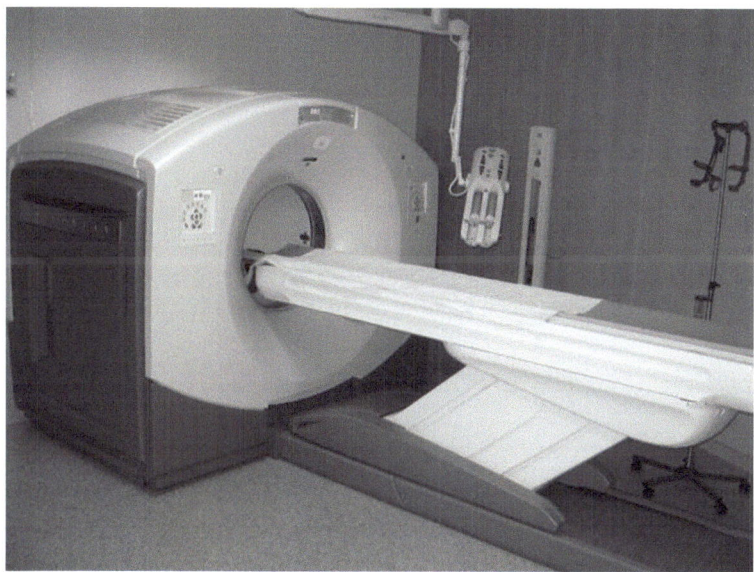

Fig. 1.8 PET/CT device

A new field of knowledge in physics has been opened up, allowing for the medical application of hadron therapy, using protons and carbon ions (hadrons). The following figure (Fig. 1.9) shows how an ion beam acts on a tumor. Conventional radiotherapy penetrates the patient's entire body and causes damage to healthy cells, while in hadron therapy the hadrons release their maximum energy at the end of their penetration track. This is illustrated in Fig. 1.9b; the curves representing the relative doses of protons and carbon ions reach a peak, known as the Bragg peak, at a depth of 15 cm and then drop off. It is possible to use the appropriate energy of hadrons so that the Bragg peak is located within the tumor, and the healthy tissue situated at higher depths is completely spared from the radiation, as is shown in Fig. 1.9a.

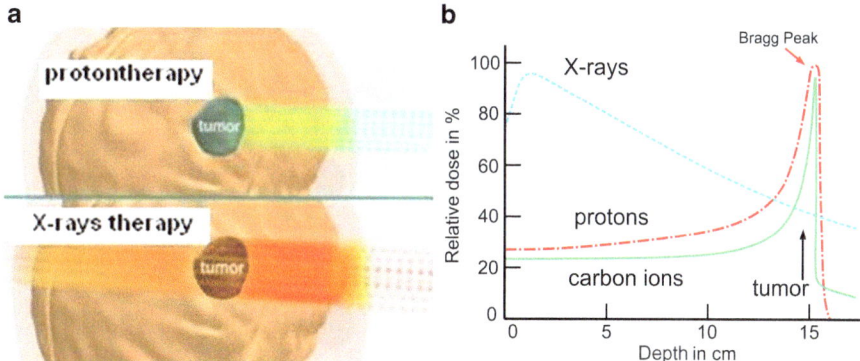

Fig. 1.9 (**a**) Behavior of proton therapy beams and X-rays in relation to the tumor. (**b**) Comparison (depth-dose) of proton and carbon ion beams with conventional radiotherapy

Certainly, hadron therapy is far superior to conventional radiotherapy, causing less damage to healthy tissue and achieving superior therapeutic results, as we will see in this Brief.

Both normal and superconducting cyclotron accelerators have been used as proton accelerators, and synchrotrons have been used to accelerate protons and carbon ions [16]. Currently, we have superconducting cyclotrons associated with high frequency linacs; these accelerate and rapidly cycle carbon ions. They are called cyclinacs [3]. Other types of accelerators, such as the DWA and laser [4], will be described in due course.

The cyclotron requires the particle to describe a spiral orbit with progressive acceleration, which is described in more detail in Chap. 3 (Fig. 1.10).

Fig. 1.10   Commercial cyclotrons (IBA and Varian/Accel)

The synchrotron imposes a circular path on the particle (Fig. 1.11).

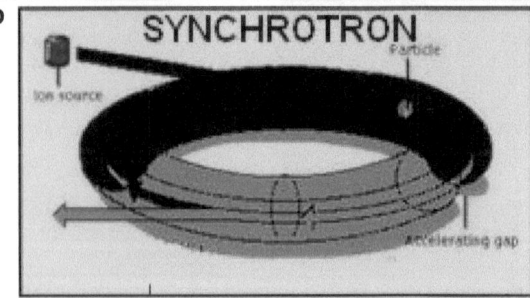

Fig. 1.11   (**a**) Commercial synchrotrons (Hitachi and Mitsubishi). (**b**) Carbon ions accelerated in a circular path

Out of a population of ten million people, about 20,000 are treated with radiation annually and resistant tumors are treated with carbon ions.

1.1.1   Addressing Hadron Therapy

I would like to clarify that we use the term "hadron therapy" as two separate words to emphasize the fact that it is a therapy involving the use of hadrons, following the recommendation of CERN; however, the simple word "hadrontherapy" has been preferred because it is more natural. Professor Ugo Amaldi [2] has suggested using the term hadrontherapy ("hadronthérapie" in French, "hadronentherapie" in German and "adroterapia" in Italian). Hadron therapy has become an important modality in cancer treatment.

Cancer can be defined as the uncontrolled growth and proliferation of a group of cells. In 1982, in the countries that were part of the European Community, 1.2 million new incidences of cancer were diagnosed. Three years later, 750,000 deaths were attributed to cancer, which amounted to the death of 1 in 5 cancer patients. In developed countries, approximately 30 % of the population suffers from cancer and about half eventually die of this disease. This corresponds to about one million deaths per year due to malignant tumors. The prognosis for individual cases varies greatly and depends on the tumor type, stage, diagnosis, general health of the patient, and other factors. In Europe, 45 % of patients experience symptom free survival for a period >5 years.

The four main approaches in treating a malignant tumor are:

1. Surgery: Direct removal of tissues affected by the cancer. This is an invasive method and is not always possible to carry out. *Surgery accounts for 22% of treatment success.*
2. Chemotherapy: Administration of drugs that prevent mitosis and cause cell death (apoptosis). *Chemotherapy has severe side effects due to the nonspecific action of the drugs in cells.*
3. Immunotherapy: Treatment of disease by inducing, enhancing or suppressing an immune response. *Immunotherapy uses the body's own immune system to help fight cancer.*
4. Radiation therapy: Tumor cells are killed by the deposition of energy. *Radiation therapy has side effects because it damages healthy tissues.* It can be administered externally using types of radiation that include the following:

   Photons: The most widely used method of energy deposition.
   Protons and ions: The method of tomorrow. Hadron therapy (proton and carbon ion therapy) is the new frontier of cancer radiation therapy.

In 1903, William H. Bragg [12] observed a peak in the energy loss of alpha particles when they passed through matter. Forty three years later, Robert Wilson [17] proposed the use of this peak for cancer therapy.

Hadron therapy is a type of radiation therapy that involves the use of highly interactive particles, which are called hadrons. There are two kinds of hadrons [6] that are classified according to their spins: baryons (spin of 1/2) and mesons (integer spin). The hadrons that we know of so far are protons, neutrons and a meson (as the only pion). The pions are the most common types of particles produced in a collision of particles and can be considered as a light proton with a mass of approximately 15 % of that of the proton. The hadrons penetrate deeper in tissue than the electron and photon, but are not as deeply penetrating as muons and neutrinos. In summary, hadrons consist of baryons (proton and neutron) and mesons (pion π^+, π^- and π°) [Nunes, M.A. (2013) Large Hadron Collider—New Era of Discoveries, in press]. In general, hadrons are considered to consist of protons, neutrons, pions and ions (alpha, carbon and neon). The strength of hadron therapy lies in the radiobiological and physical properties of the hadron particles; they can penetrate tissues with limited diffusion and release their maximum energy shortly before stopping. These characteristics enable the precise definition of the region to be specifically irradiated. Hadron therapy allows radiation delivery that is better controlled than in conventional radiation therapy; the radiation beam can follow the movement of the tumor by means of "gating" and the "intensity-controlled raster scan method" during treatment, as we will see later. Thus, using hadrons, the tumor can be irradiated with less damage to healthy tissue than is the case using conventional X-rays [1, 5].

In the USA, only proton therapy is used; therefore, I have provided some relevant information regarding the proton. Carbon ion therapy has flourished in Japan and also in Europe, but the original concept was American and was first developed at Berkeley University in California.

Electrons were discovered at the end of the nineteenth century by Joseph J. Thomson. Thomson proved experimentally the existence of corpuscles (later called electrons) in cathode rays, after the passage of an electric current through a vacuum tube. The popular model for the atom at the time was the famous Thomson's "plum pudding" model. It consisted of a spherical mass with a positive charge in which the negatively charged electrons were soaked. However, in 1911, the British physicist Ernest Rutherford (a student of Joseph J. Thomson) and colleagues used alpha particles as projectiles to bombard the atoms in thin sheets of gold. They did not know the composition of these alpha particles (helium nuclei with two protons and two neutrons). The source of these particles was radioactive polonium. They were emitted in all directions, but Rutherford was interested only in those that hit the target (they were therefore collimated). After traversing the target (a plate of gold), the particles were detected using a moving screen coated with zinc sulfide, which flashed when hit by an alpha particle. Thus, they could study the position of the emerging alpha particles. The results were simply fantastic. Most particles freely crossed the bulwark (empty lots) and did not suffer even a slight deflection. Some of the alpha particles returned as if they had collided with something massive, which was inconsistent with the pudding model mentioned earlier. A new model for the atom was proposed: the atom contained a solid core in which all the mass and all of the positive charge was concentrated. Alpha particles collide with these nuclei. To balance the positive charges of the nuclei, electrons exist in the outer region beyond the nucleus. Rutherford came to determine the nature of alpha particles with the following method. Alpha particles were collected within a vacuum container. After compression of a rarefied gas, he passed an electrical discharge through the system and observed the spectrum of the emitted light, which represented the characteristics of helium; indeed, the helium atoms that were ionized combined with the electrons in the evacuated container and formed helium gas. The experiments with alpha particles were continued by Rutherford and James Chadwick. They started bombarding nitrogen nuclei with alpha particles. The alpha particles passed straight through and appeared as hydrogen nuclei. Therefore, it was accepted that the nitrogen nuclei were formed from hydrogen nuclei, and that perhaps all nuclei were formed from hydrogen nuclei; the hydrogen nucleus was given the name "proton". Protons had a unit of electrical charge that was equal and opposite to that of the electron. Taking the mass of the proton as a unit, the electron mass would be 1/1,800 the mass of the proton. Thus, the core would consist of protons. It was a simple system. Rutherford, amazed by the results of his experiments, wrote: *"It was quite the most incredible event that ever happened to me in my life. It was as incredible as if you fired a 15-inch shell at a piece of tissue paper and it came back and hit you. On consideration, I realized that this scattering backwards must be the result of a single collision, and when I made calculations I saw that it was impossible to get anything of that order of magnitude unless you took a system in which the mass of the atom was concentrated in a minute nucleus."* In 1911, Rutherford proposed that the atom had a very small mass at its center, maintaining all the positive charge required to balance the negative charge of all the electrons circling around the positively charged center (core). This was the first time the correct structure had been proposed for the atom. Rutherford was the creator of nuclear physics and for his work he received the Nobel Prize in Chemistry. Rutherford was not happy with his placement in the chemistry section by the selection committee, since he was a physicist. At the location where Rutherford and colleagues performed their experiments with radioactive polonium, a large building was built; this had to be closed after four people and two other researchers died, all diagnosed with cancer.

In the 1950s, the structure of the nucleus was systematically investigated using electron scattering. Robert Hofstadter was awarded the Nobel Prize in Physics in 1961 for his pioneering studies on electron scattering. The investigation of protons and neutrons as the building blocks of nature was intensified during this period and a large number of particles called hadrons were discovered. However, it was Murray Gell-Mann, the Nobel Prize winner in Physics in 1969, who successfully solved the problem of hadrons; the various types of hadrons were found to be related and behaved as members of a kind of family with the introduction of building blocks called "quarks". All known hadrons could be constructed from three quarks and their anti-particles. It was a great concept and the conceptual simplification involving the quark was immediately accepted. The intellectual courage of Gell-Mann for introducing unseen particles with a non-integer charge must be emphasized. Other researchers, such as Wolfgang Pauli, were also important. However, now the existence of the none-free quark is accepted. Should it be only mathematical quantities that are included in the equations used in physics? The elementary particles initially had a history involving Greek names, but use of the term quark followed the trend of modern physics in using strange names. It seems that the word quark stemmed from comic fiction (Finnegans Wake by James Joyce—Three quarks for Muster!). It may be that quarks are composed of even smaller entities. Perhaps the LHC will make some contributions in this direction.

In 1964, Murray Gell-Mann and George Zweig independently proposed a theory to explain all hadrons. Gell-Mann proposed the term quark for the new particles. The model has worked very well, but no experiment has detected a free quark to date. However, there is strong evidence for the existence of quarks inside protons, neutrons and other particles. In 1970, neutrinos and electrons were used to investigate the role of protons in the formation of alpha particles, similar to those that were used in the experiments of Rutherford. These experiments showed that the proton consists of three particles, each with spin 1/2, with two charges of +2/3 and one with a charge of −1/3.

James D. Bjorken and Richard P. Feynman (Nobel Prize winner for Physics in 1965), both theoretical physicists, have shown that electron scattering at large angles can be explained by the existence of hard granules, namely quarks, in nuclei. However, their findings could not be explained based solely on quarks. Therefore, it has been suggested that there are neutral components in the nuclei, called "gluons," which are the intermediates of strong force. A new era in the history of physics was being introduced.

1.2 Hadron Therapy Timeline

Now, I will look at some historical data on hadron therapy and their timeline:

1. The first cancer treatment of deep-seated tumors using radiation (X-rays) was carried out by the brothers Lawrence (Ernest O. Lawrence and John Lawrence) in 1937. The treatment seemed to have cured the uterine cancer (inoperable) of their mother, but the disease was probably misdiagnosed.
2. JS Stone and John Lawrence, both medical doctors, used neutron therapy in patients starting in 1938, with a program that involved 250 patients. Stone concluded that neutron therapy was a "delayed stressful" modality and "should not be continued". No additional work was undertaken regarding neutron therapy over the following 25 years.
3. Siemens and Varian of the USA built the first X-ray linacs.
4. Most patients are treated using X-rays. There are 10,000 linacs worldwide and they are used to treat 4,000,000 patients per year.
5. Hadron therapy (Bragg peak) was first suggested by Robert Wilson [17]. Berkeley and Harvard Universities were pioneers in this area.
6. The combination of the Bevatron particle accelerator with the SuperHILAC (linear accelerator used as an injector for heavy ions) was named the Bevalac. During the 1970s, the use of heavy ions was carefully developed in the Bevalac (the only accelerator capable of accelerating any nucleus in the periodic table to relativistic energies) from basic biology to patient treatment. Great effort was expended on research and development to answer such questions involving

which types of cancer responded best to treatment and the optimal radiation doses. Many scientists, including Joe Castro, Bill Chu, John Lyman, Cornelius Tobias, Eleanor Blakely, Ted Philips and others, participated in these studies. The Bevalac was used two thirds of the time for medical studies and one third of the time for nuclear physics studies. The ground work was laid for the relativistic heavy ion collider (RHIC) and the LHC.

7. Based on the work carried out at Berkeley, the heavy ion medical accelerator in Chiba (HIMAC) was built in Chiba, Japan. It was the first facility dedicated to the treatment of cancer using ions. Although none of these accelerators were established in the USA, many were built in Japan (approximately 50) and some were also built in Europe.

8. Therapy using pions and neutrons has been used in the past, but did not prove to be of great interest to the oncology community, although treatment using fast neutrons was initiated at Fermilab in the USA.

Timeline: *key events*

1930: Ernest Lawrence invents the cyclotron accelerator.

1938: Neutron therapy is developed by John Lawrence and JS Stone at Berkeley University.

1946: For the first time, Robert Wilson suggests that energetic protons could be an effective cancer treatment method in a study published in 1946, while he was involved in designing the Harvard Cyclotron Laboratory.

1948: Extensive studies in Berkeley confirm Wilson's suggestion.

1954: Protons are used to treat patients in Berkeley.

1957: Uppsala University, Sweden duplicates the results obtained at Berkeley.

1961: The Harvard Cyclotron Laboratory teams up with Massachusetts General Hospital to use proton therapy. The first treatment takes place at Harvard. Over a period lasting until 2002, 9,111 patients were treated. The hospital was closed in 2002.

1968: Installation of a particle accelerator takes place in Dubna, Russia.

1969: Installation of a proton accelerator in Moscow is completed.

1970: The Massachusetts General Hospital conducts the first study regarding radiotherapy with protons/photons for the treatment of prostate cancer.

1972: Fast neutron therapy is initiated at the MD Anderson Hospital in Texas (soon more units open at six locations in the USA).

1974: Patients are treated with a pi-meson beam at the Los Alamos National Laboratory in New Mexico. Treatment is terminated at the end of 1981; It was started and finished at the Paul Scherrer Institute (PSI) in Switzerland and the TRIUMF facility in Canada.

1975: A proton therapy facility is opened in St. Petersburg, Russia.

1975: A team at Harvard University pioneer the treatment of eye cancer with protons.

1976: Fast neutron therapy is started at Fermilab. This facility is closed in 2003 after treating 3,100 patients.

1977: Cancer treatment with ions is initiated using the Bevalac. The facility is closed in 1992, after treating 233 patients.

1979: Proton therapy commences in Chiba, Japan, using the HIMAC.

1980: The design and construction of the first facility dedicated to clinical proton therapy treatments takes place at Loma Linda University Medical Center in California.

1988: Proton therapy is approved by the Food and Drug Administration (FDA) in the USA.

1989: Proton therapy commences using the Clatterbridge accelerator in the UK.

1990: Medical plans covering proton therapy are developed by the Particle Therapy Co-Operative Group (PTCOG): www.ptcog.web.psi.ch

1990: The first hospital-based proton treatment facility is opened in Loma Linda, California.

1991: Proton therapy is initiated in Nice and Orsay in France.

1992: The Berkeley cyclotron accelerator is closed after treating more than 2,500 patients.

1993: Proton therapy is initiated in Cape Town, South Africa.

1993: Indiana University Health Proton Therapy Center in the USA treats its first patient with protons.

1994: Therapy with carbon ions is started at the HIMAC in Japan. By 2008, more than 3,000 patients have been treated.

1996: Installation of a proton accelerator at the PSI in Switzerland.

1998: Installation of a proton accelerator in Berlin, Germany.

1990–2000: More than 25,000 patients are treated worldwide with proton therapy.

2001: Massachusetts General Hospital in the USA opens a proton therapy center.

2006: The MD Anderson Hospital in Texas opens a proton therapy center.

2007: A proton therapy center is opened in Unity, Jacksonville, Florida.

2008: Neutron therapy is restarted at Fermilab.

Only the USA, Europe, Asia and Africa have these hadron treatment resources. We have no hadron therapy facilities in South America; there is only one ongoing project in Argentina that involves a related technology known as boron neutron capture therapy (BNCT). Locations of hadron therapy facilities around the world are detailed below; however, the latest updates are available at the website (ptcog. web.psi.ch).

- In the USA:

  - Loma Linda, CA (1990); Boston, MA (2001); Bloomington, IN (2004): Houston, TX (2006); Jacksonville, FL (2006). (Ion therapy formally took place at the Berkeley Lab, CA)

- In the rest of the world:

  - Japan: Chiba (1994); Kashiwa (1998); Tsukuba (2001); Hyogo (2001); Wakasa (2002): Shizuoka (2003), Tsunuga.
  - Germany: Munich; Essen; Heidelberg (HIT); Marburg; Kiel.
  - Europe: Pavia, Italy; Orsay, France; Trento, Italy; Uppsala, Sweden; Vienna, Austria; Lyon, France; Paul Scherrer Institute, Switzerland (1984); St Petersburg, Russia; Moscow, Russia; Dubna, Russia.
  - Other Places: Seoul, Korea; Zibo, China (2004).

Hadron therapy was initiated at Berkeley in 1938 [18]. It developed rapidly, and there are now advanced centers of physics such as CERN in Geneva, which has the largest hadron accelerator in the world (the LHC). CERN also has the best school of hadron therapy in the world, with scholars attending from around the world, mainly ENTERVISION research fellows. Approximately one third of the 15,000 accelerators operating in the world are used in medicine, and 3 % are employed in nuclear medicine and 30 % in radiation therapy. Most of them produce X-rays, while only 25 are used as beam sources of hadrons.

Tumors that are more sensitive to treatment using hadron therapy include chondrosarcomas, arteriovenous malformations and uveal melanoma. It is important that a multidisciplinary team, consisting of medics, physicists, technicians, and others is involved in research and treatment so that there is an efficient exchange of ideas. Hadron equipment is very expensive and this is perhaps the most important limitation in acquiring this resource throughout the world.

Two corroborative pictures regarding the treatment efficiency of fast neutron therapy involving an inoperable tumor are provided in Fig. 1.12.

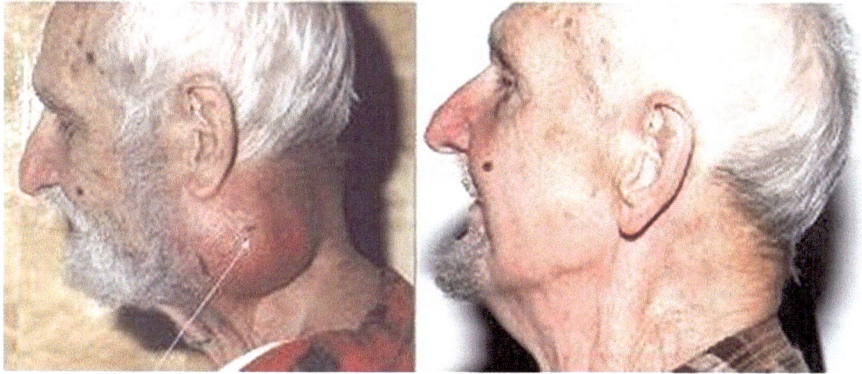

Fig. 1.12 Inoperable squamous cell tumor, before and 2 years after fast neutron therapy [15]. Arlene Lennox (presented in a seminar at Fermilab on Nov 21, 2003)

The neutron therapy unit at Wayne State University in Detroit, USA, called the Gershenson Radiation Oncology Center at Karmanos Center (KCC/WSU), has more experience than any other facility in the world regarding the use of fast neutron therapy for prostate cancer; they have treated >1,000 patients over the last 10 years. The KCC produces its neutron beam by accelerating deuterons with an energy of 48.5 MeV on to a beryllium target. The deuterons are accelerated using a superconducting cyclotron, where the source of neutrons can spin 360° around the patient. Neutrons can be obtained by accelerating protons (p) or deuterium (^{2}H) and making them collide with a beryllium (Be) or lithium (Li) target provoking reactions of the type: ^{9}Be (p, n) ^{9}B; ^{7}Li (p, n) ^{7}Be; and ^{3}H (^{2}H, n) ^{4}He.

References

1. Alonso JR (2000) Review of ion beam therapy: present and future. In: Proceedings of EPAC, Vienna, Austria, p 235
2. Amaldi U (2007) Hadrontherapy: applications of accelerator technologies to cancer treatment. TERA Foundation conference presentation (17 May 2007). Available from: http://basroc.rl.ac. uk/basroc_files/icpt/.../RAL-Amaldi-17.5.07.pdf
3. Amaldi U, Braccini S, Citterio A et al (2009) Cyclinacs: fast-cycling accelerators for hadron-therapy. arXiv:0902.3533 (physics.med-ph)
4. Amaldi U et al (2010) Accelerators for hadrontherapy: from Lawrence cyclotrons to linacs. Nucl Instrum Methods Phys Res A 620:563–577
5. DeLaney TF, Kooy HM (eds) (2008) Proton and charged particle radiotherapy. Lippincott Williams & Wilkins, Philadelphia, PA
6. Lincoln D (2009) The Quantum Frontier: the large Hadron Collider. Johns Hopkins University Press, Baltimore, MD
7. http://www.bookrags.com/biography/william-d-coolidge-woc/
8. http://www.nobelprize.org/nobel_prizes/chemistry/laureates/1935/joliot-curie-bio.html
9. http://www.nobelprize.org/nobel_prizes/physics/laureates/1901/rontgen-bio.html
10. http://www.nobelprize.org/nobel_prizes/physics/laureates/1903/becquerel-bio.html
11. http://www.nobelprize.org/nobel_prizes/physics/laureates/1903/marie-curie-bio.html
12. http://www.nobelprize.org/nobel_prizes/physics/laureates/1915/wh-bragg-bio.html
13. http://www.nobelprize.org/nobel_prizes/physics/laureates/1938/fermi-bio.html
14. http://www.nobelprize.org/nobel_prizes/physics/laureates/1939/lawrence-bio.html
15. http://www-bd.fnal.gov/ntf/reference/hadrontreat.pdf
16. Ma CC-M, Lomax T (eds) (2012) Proton and carbon ion therapy, Imaging in medical diagnosis and therapy. CRC Press, Boca Raton, FL
17. Wilson R (1946) Radiological use of fast protons. Radiology 47:487–491
18. www.slac.stanford.edu/slac/sass/talks/aiden_sass.pdf

Chapter 2
Hadron Therapy

2.1 Mechanism of Action of Hadron Therapy at the Molecular Level

The cells in the body replicate and die. When replicating, they pass their deoxyribonucleic acid (DNA) to daughter cells. DNA encodes the genetic instructions used in the development and functioning of all known living organisms. This genetic information is encoded as a sequence of nucleotides namely adenine, thymine, cytosine and guanine. DNA has a double-helix structure consisting of two strands of nucleotides that are connected by hydrogen bonds. The DNA backbone is resistant to cleavage and the double-stranded structure provides molecules with a built-in duplicate of the encoded information. DNA is organized into long structures called chromosomes. The DNA determines which proteins the cells can produce. Cells with functional DNA damage do not produce viable daughter cells. Radiation therapy works by damaging the DNA. We know that the DNA in existence 3.5 billion years ago had to deal with cosmic rays and other adverse factors. Thus, it was subjected to natural selection. When one of the strands of DNA is damaged, another strand can repair it. In some cases the DNA can be used from another chromosome (Fig. 2.1).

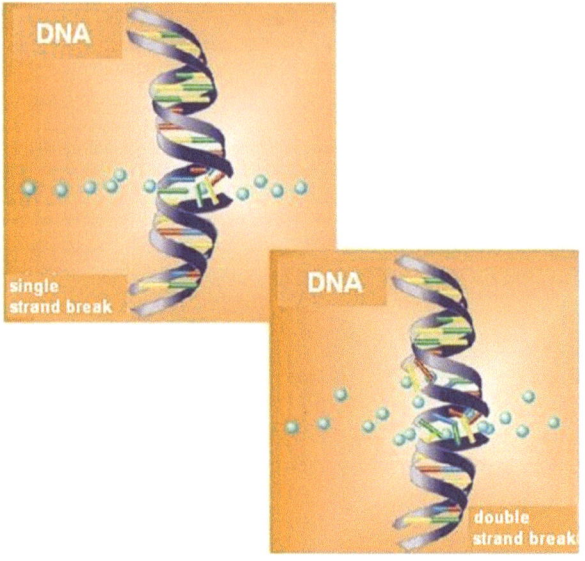

Fig. 2.1 Breaks in single and double stranded DNA

Bombardment with heavy ions is much more effective in damaging both strands of DNA than X-rays because they hit the target more precisely and produce more consistent damage [2, 6, 13] (Fig. 2.2).

Fig. 2.2 Damage to DNA caused by X-rays and heavy ions

For the treatment of tumors that are located more deeply in body tissue, the proton accelerator must produce beams with higher energy expressed in electron volts (eV). Tumors closer to the body surface are treated using low-energy protons. The accelerators used in hadron therapy typically produce protons with energies in the range 70–250 MeV. Tumor cell damage can be maximized by adjusting the energy of the protons during treatment. Tissues nearer the body surface (outside of the tumor volume) experience less radiation damage and deep tissues within the body are exposed to a low number of protons; thus, their radiation dose exposure is very small. The intrinsic accuracy of a hadron beam in terms of the dose delivered to tumor tissue is the main advantage of hadron therapy, as compared with traditional therapy involving photons and electrons.

When a charged particle moves through a medium, it ionizes atoms in its path and delivers a dose of radiation. As mentioned previously, when plotting the energy of this ionizing radiation as the particle travels through matter, there is a sharp peak called Bragg peak, which is seen with protons, alpha-particles and ions immediately before they come to rest. Therefore, the energy peak occurs because the interaction cross section of the particle increases with decreasing energy. This phenomenon is used in hadron therapy to concentrate the effect of the beam on the tumor being treated, while minimizing the adverse effects on the surrounding healthy tissue.

Figure 2.3 shows how X-ray and ion beams penetrate human tissue.

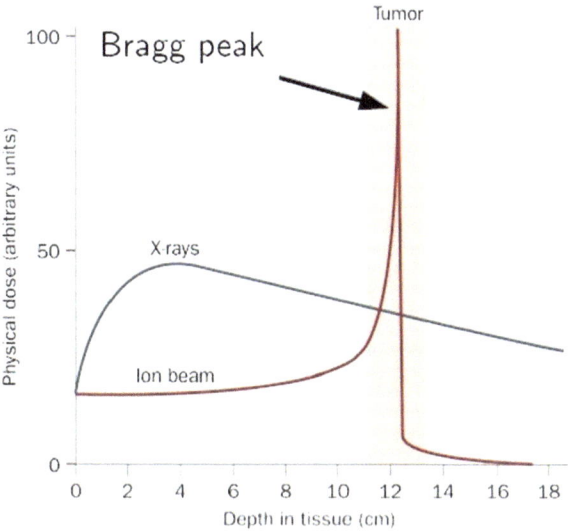

Fig. 2.3 The Bragg peak

For protons and heavy ions, the radiation dose increases as the particles penetrate through tissue and continuously lose energy (i.e., the dose increases with increasing tissue thickness until the Bragg peak is reached), which occurs near the end of the travel range of the particle. Note that beyond the Bragg peak the dose falls to zero for protons and nearly zero for heavy ions. The figure also shows that X-rays penetrate deeper, but the radiation dose absorbed by the tissue shows an exponential decay with increasing thickness.

2.2   Strengthening Concepts Regarding Variations in Radiation Therapy

Radiation therapy has been indispensable in treating cancer, in particular X-rays and gamma rays (photonic therapy). This type of radiation removes electrons from the atoms of tumor cells and destroys cellular DNA, and consequently the construction plans for essential proteins. Thus, cells stop replicating and die, as mentioned earlier. However, it must be emphasized that energy is only partially transferred to the tumor. Modern techniques have partially solved this problem [4]. Photon beams impact the tumor from multiple directions and accurately discharge their maximum energy. At the same time, mobile apertures sweep healthy tissues sensitive to radiation. This approach is called intensity modulated radiation therapy (IMRT); it improves cancer treatment by precise tumor targeting using conventional X-rays.

Hadron therapy is a non-conventional type of radiation therapy. Ion radiation beams consist of positively charged ions, namely ions that have been obtained from atomic nuclei that have lost their electrons. In general, the cores of atomic hydrogen (protons) and the heavy atomic nuclei of carbon (heavy ions) are used. Atomic nuclei are accelerated to approximately 3/4 of the speed of light before they reach the tumor. The depth of penetration can be increased by increasing the speed of the ions. When the ions impact the body, they travel very quickly through the outer layers of tissue before the final deceleration at their ultimate penetration depth, where they rapidly lose their energy. Thus, the ion beam is exactly what is needed to treat deep seated tumors. Tumors with irregular borders can be scanned using the intensity-controlled raster scan method.

Figure 2.3 shows the maximum radiation dose that can be achieved by photons; this reaches a maximum in healthy normal tissue overlying the tumor. In contrast, the ion beam specifically targets the tumor, releasing all of its damaging energy therein.

Figure 2.4 shows the different distributions of the biologically effective dose for different types of radiation.

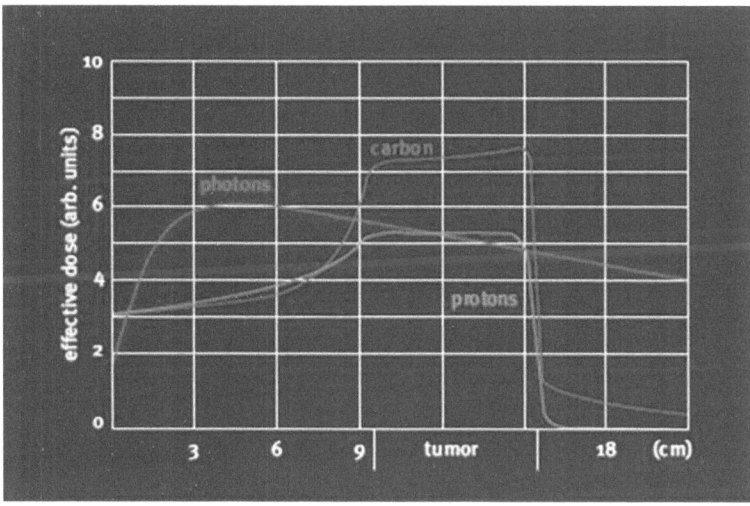

Fig. 2.4 Differences in the biologically effective dose profiles of photons, protons and carbon ions with tissue depth. Courtesy of HIT

Thus, in the case of photon therapy the maximum radiation dose is delivered before reaching the tumor, and decreases exponentially with increasing tissue depth. As for proton and carbon ion therapy, the radiation dose delivered is highest in the tumor and falls rapidly outside of the tumor. Carbon ions are much more biologically effective than protons, as can be seen in Fig. 2.4 [5]. Note that the flattening of the Bragg peak is reached within the tumor using the intensity-controlled raster scan method. This procedure is essential for distributing all of the damaging radiation throughout the tumor volume and not just at one point. In a cell, the carbon ion releases about 24 times more energy than a proton with the same range. This produces a dense column of ionization, especially near the Bragg peak, causing many double strand breaks and damage to multiple sites when traversing the DNA in the cell nucleus.

Proton beams with an energy of 200–250 MeV and a low current of 2 nA, and carbon ion beams with an energy of 350–450 MeV and currents of approximately 0.2 nA, are useful in treating tumors because of five physical properties. The first is that these beams release their energy maximum at the end of their range (the Bragg peak). Second, these beams penetrate tissue with very little diffusion, and carbon ions are three to four times better in this regard than protons [8]. Third, electromagnets can be used to precisely focus the beams and brush scan them at variable depths of penetration because the particles are charged; this allows any part of the tumor to be irradiated. The fourth property concerns the relative biological effectiveness (RBE) and relates to carbon ions in particular; they release 24 times more energy than protons, producing greater damage to the DNA in the cell nucleus. This involves multiple double strand breaks that are not amenable to repair by the usual cellular mechanisms. Thus, the effects are qualitatively different from those produced by other types of radiation and therefore the carbon ions can exert greater control of radioresistant tumors than protons and X-rays. Fifth, this therapy causes negligible damage to the normal tissue surrounding the tumor and does not constitute a "frying-pan" effect (one patient's expression) as is the case with conventional radiotherapy.

2.3   Which Conditions Can Be Treated Using Hadron Therapy?

The following conditions can be treated using hadron therapy:

1. Cells contained in a solid tumor, without metastasis. Prostate tumor is an example for which good therapeutic results have been achieved.
2. Brain and spinal cord tumors, and isolated metastasis in the brain.
3. Arteriovenous malformations without tumor involvement.
4. Eye diseases, and inoperable tumors in the neck and base of the brain.
5. Lung tumors.
6. Radioresistant tumors and tumors of the salivary gland, liver and pancreas.
7. Pediatric tumors with excellent therapeutic results.

2.4   Hospitals and Research Centers Treating Cancer with Hadrons

Proton and ion therapy	
North America	• Loma Linda Proton Treatment Center at Loma Linda University Medical Center (LLUMC), California. A 250 MeV proton synchrotron with a pulse width of 300 ms and a frequency of 2.2 s
	• Eye Therapy at the Crocker Nuclear Laboratory, University of California, Davis, including the LBL eye cancer therapy program
	• The Northeast Proton Therapy Center (NPTC) at Massachusetts General Hospital. See the Harvard Cyclotron Laboratory (HCL) at Harvard University, and the home page of the Particles newsletter, sponsored by the Proton Therapy Co-operative Group (PTCOG)
	• Midwest Proton Radiation Institute at the Indiana University Cyclotron Facility. Up to 210 MeV proton cyclotron with 0.5–1 μA current
	• Cancer treatment with protons at TRIUMF National Laboratory, Canada

(continued)

(continued)

Proton and ion therapy

Europe	• Centro Nazionale di Adroterapia Oncologica (CNAO), Italy
	• Radiation Medicine at the Paul Scherrer Institute (PSI), Villigen, Switzerland. A running proton cyclotron with an active spot scanning system
	• Heavy Ion Cancer Therapy at Gesellschaft für Schwerionenforschung mbH (GSI), Darmstadt, Germany. Running a synchrotron proton/ion accelerator with a raster scanning system
	• Proton therapy project at Kernfysisch Versneller Instituut (KVI), Groningen, The Netherlands. A running proton cyclotron with a passive scanning system
	• Clatterbridge Centre for Oncology in the UK is under construction. According to their website it is due to open in 2017
	• The Center of Prótontherapy of Orsay (CPO) near Paris, France. A running proton synchrocyclotron with a passive scanning system. An available basic introduction to proton therapy and related topics and several useful links
	• Centre Antoine Lacassagne (CAL), Nice, France. A running MEDICYC 65 MeV proton cyclotron for proton and neutron therapy
	• Department of Oncology of Uppsala University Hospital, Sweden and the Svedberg Laboratory (TSL) at Uppsala University have a proton therapy research program
	• Med-AUSTRON part of the Neutron Spallation Source Project AUSTRON, Austria. See also AUSTRON
	• The Prague Medical Synchrotron Project, a joint proton therapy project of the Nuclear Research Institute in Rez and the First Faculty of Medicine of Charles University, Czech Republic
	• Medical Physics Department of ITEP at Dubna, Russia
Africa	• Medical Radiation Group of the National Accelerator Center (NAC) in South Africa
Asia	• Proton Medical Research Center (PMRC) at the University of Tsukuba. It uses the 500 MeV booster synchrotron of KEK operated at 20 Hz. A new facility is under development
	• National Institute of Radiological Sciences (NIRS) Chiba, Japan. They have a research center for charged particle therapy

Neutron/boron neutron capture therapy

Neutrons with energies of up to 50 MeV are used to treat specific types of cancer

Boron neutron capture therapy (BNCT) is an experimental approach to cancer treatment that combines boron and low-energy neutrons. Centers that have evaluated this modality include:

• Medical Department, Brookhaven National Laboratory, USA
• Lawrence Berkeley National Laboratory, USA
• Fermilab Neutron Therapy Facility, USA
• UCLA, USA
• The INEEL/MSU Center for Advanced Radiation Therapies at Montana State University, USA
• Kyoto University Research Reactor, Japan

The following gives a current listing of the particle therapy facilities currently in operation together with patient statistics. The Particle Therapy Co-Operative Group (PTCOG), which provided the data in Table 2.1, has made the data available for the benefit of patients requiring hadron therapy. The PTCOG reports on data concerning radiotherapy with protons, light ions and heavy particles.

Visit the site http://ptcog.web.psi.ch for more details.

Table 2.1 Hadron therapy facilities worldwide. Last update: 10-April-2013

Where	Country	Particle	S/C[a], max. energy (MeV)/Beam direction		Start of treatment	Total number of patients treated	Date of total
ITEP, Moscow	Russia	p	S 250	1 horiz.	1969	4,246	Dec-10
St. Petersburg	Russia	p	S 1000	1 horiz.	1975	1,386	Dec-12
PSI, Villigen	Switzerland	p	C 250	1 gantry[b], 1 horiz.	1996	1,409	Dec-12
Dubna	Russia	p	C 200[c]	1 horiz.	1999	922	Dec-12
Uppsala	Sweden	p	C 200	1 horiz.	1989	1,267	Dec-12
Clatterbridge	England	p	C 62	1 horiz.	1989	2,297	Dec-12
Loma Linda	CA, USA	p	S 250	3 gantry, 1 horiz.	1990	16,884	Dec-12
Nice	France	p	C 65	1 horiz.	1991	4,692	Dec-12
Orsay	France	p	C 230	1 gantry, 2 horiz.	1991	5,949	Dec-12
NRF—iThemba Labs	South Africa	p	C 200	1 horiz.	1993	521	Dec-11
IU Health PTC, Bloomington	IN, USA	p	C 200	2 gantry[d], 1 horiz.	2004	1,688	Dec-12
UCSF	CA, USA	p	C 60	1 horiz.	1994	1,515	Dec-12
HIMAC, Chiba	Japan	C-ion	S 800/u	horiz.[d], vertical[d]	1994	7,331	Jan-13
TRIUMF, Vancouver	Canada	p	C 72	1 horiz.	1995	170	Dec-12
HZB (HMI), Berlin	Germany	p	C 72	1 horiz.	1998	2,084	Dec-12
NCC, Kashiwa	Japan	p	C 235	2 gantry[d]	1998	1,226	Mar-13
HIBMC, Hyogo	Japan	p	S 230	1 gantry	2001	3,198	Dec-11
HIBMC, Hyogo	Japan	C-ion	S 320/u	horiz., vertical	2002	788	Dec-11
PMRC (2), Tsukuba	Japan	p	S 250	2 gantry	2001	2,516	Dec-12
NPTC, MGH Boston	MA,USA	p	C 235	2 gantry[d], 1 horiz.	2001	6,550	Oct-12
INFN-LNS, Catania	Italy	p	C 60	1 horiz.	2002	293	Nov-12
SCC, Shizuoka Cancer Center	Japan	p	S 235	3 gantry, 1 horiz.	2003	1,365	Dec-12
STPTC, Koriyama	Japan	p	S 235	2 gantry, 1 horiz.	2008	1,812	Dec-12
WPTC, Zibo	China	p	C 230	2 gantry, 1 horiz.	2004	1,078	Dec-12
MD Anderson Cancer Center, Houston	TX, USA	p	S 250	3 gantry[d], 1 horiz.	2006	3,909	Dec-12
UFPTI, Jacksonville	FL, USA	p	C 230	3 gantry, 1 horiz.	2006	4,272	Dec-12

NCC, Ilsan	South Korea	p	C 230	2 gantry, 1 horiz.	2007	1,041	Dec-12
RPTC, Munich	Germany	p	C 250	4 gantry[b], 1 horiz.	2009	1,377	Dec-12
ProCure PTC, Oklahoma City	OK, USA	p	C 230	1 gantry, 1 horiz, 2 horiz/60 deg.	2009	1,045	Dec-12
HIT, Heidelberg	Germany	p	S 250	2 horiz.[b]	2009	252	Dec-12
HIT, Heidelberg	Germany	C-ion	S 430/u	2 horiz.[b]	2009	980	Dec-12
UPenn, Philadelphia	PA, USA	p	C 230	4 gantry, 1 horiz.	2010	1,100	Dec-12
GHMC, Gunma	Japan	C-ion	S 400/u	3 horiz., vertical	2010	537	Dec-12
IMP-CAS, Lanzhou	China	C-ion	S 400/u	1 horiz.	2006	194	Dec-12
CDH Proton Center, Warrenville	IL, USA	p	C 230	1 gantry, 1 horiz, 2 horiz/60°	2010	840	Dec-12
HUPTI, Hampton	VA, USA	p	C 230	4 gantry, 1 horiz.	2010	489	Dec-12
IFJ PAN, Krakow	Poland	p	C 60	1 horiz.	2011	15	Dec-12
Medipolis Medical Research Institute, Ibusuki	Japan	p	S 250	3 gantry	2011	490	Dec-12
CNAO, Pavia	Italy	p	S 250	3 horiz./1 vertical	2011	58	Mar-13
CNAO, Pavia	Italy	C-Ion	S 400/u	3 horiz./1 vertical	2012	22	Mar-13
ProCure Proton Therapy Center, Somerset	NJ, USA	p	C 230	4 gantry	2012	137	Dec-12
PTC Czech r.s.o., Prague	Czech Republic	p	C 230	3 gantry, 1 horiz.	2012	1	Dec-12
SCCA, Proton Therapy, a ProCure Center, Seattle	WA, USA	p	C 230	4 gantry	2013	1	Mar-13

[a]S/C Synchrotron (S) or Cyclotron (C)
[b]With beam scanning
[c]Degraded beam
[d]With spread beam and beam scanning

2.5   Particle Therapy Facilities in Operation (Including Patient Statistics)

Patient statistics are for particle therapy facilities worldwide. They include the number of patients treated in facilities that are presently in operation and in facilities that are now closed. The data were received from centers up until the end of 2012 (pdf-file for download prepared by PTCOG Secretary).

An individual hadron therapy treatment costs approximately 70,000 US dollars. The equipment costs 80 million US dollars, 70 % of which is spent on the facilities. It is easy to see why there are no hadron therapy facilities in South America. Fortunately, in South Africa, they have such equipment. Because the objective is treating human diseases, such prices should not be an obstacle. Figure 2.5 shows the appearance of typical hadron equipment. In this example, the machine has a mean radius of 5×10 m and operates at an energy of 200–250 MeV.

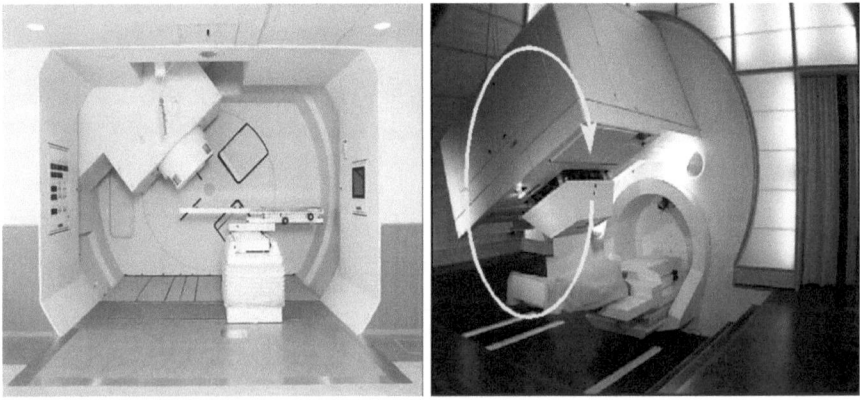

Fig. 2.5 Commercially available hadron equipment. The *left panel* is an image of an isocentric gantry manufactured by Mitsubishi and the *right panel* is an image of an eccentric gantry made by PSI

2.6   Tumor Localization

Certainly, rigorous planning is essential to ensure treatment with the lowest possible damage to healthy normal tissue. Currently, we use advanced techniques for the administration of conventional radiotherapy in Brazil, such as intensity modulated radiation therapy (IMRT), computed tomography (CT), magnetic resonance imaging (MRI) and positron emission tomography (PET). All of these techniques allow the exact localization of tumors and provide information concerning their physical characteristics.

PET-CT allows the use of PET image acquisition together with the imaging features offered by CT. Thus, PET-CT can not only detect tumors, but also provide data on tumor characteristics such as malignancy, recurrence and response to therapy. This modality usually employs fluorodeoxyglucose (FDG), which is a positron emitter, to indicate tumoral activity.

To detect tumoral activity in the brain, thalium-201 is used because of the greater sensitivity of brain tumors to this agent relative to FDG. The normal brain cannot accumulate thalium-201. However, caution must be taken using thalium-201 because its decay can generate mercury, which is harmful.

2.7   Treatment of Cancer Using Boron Neutron Capture Therapy

The World Health Organization (WHO) estimates that each year, ten million people are diagnosed with cancer, and 60 % of these die. The WHO also predicts that within 20 years the number of people diagnosed with cancer will increase by 50 %. In Brazil, in 2003, approximately 400,000 new patients (about 30 %) died of cancer. The development of BNCT has relied on the dedication of numerous international organizations. The two organizations responsible for the initial development of BNCT were the Massachusetts General Hospital in Boston, MA and Brookhaven National Laboratory in Upton, NY. BNCT can be used to treat the following types of disease [9]:

1. Glioblastoma multiforme (a malignant tumor that occurs in the central nervous system).
2. Skin melanoma.
3. Multifocal liver tumors.
4. Oral cancer.
5. Undifferentiated thyroid cancer.
6. Head and neck cancer.
7. Rheumatoid arthritis.

BNCT involves the administration of a compound containing boron-10, which selectively concentrates in tumor cells. Next, the tumor is exposed to a beam of thermal neutrons at a dose that does not cause damage to the healthy tissue surrounding the tumor. Boron-10 has a high capture cross section for thermal neutrons, hence this element is used. The low energy neutrons interact with the boron-10 in a neutron capture reaction to yield high linear energy transfer alpha particles, recoiling lithium-7 nuclei and gamma rays. These different types of radiation have short-range effects that are restricted to the cancer cells containing the boron-10.

BNCT has encountered a number of problems because the original low energy neutron beams did not penetrate deeply enough in tissue and because of low boron-10 accumulation in tumors. In the 1990s, MIT researchers performed studies for 2 years on patients with malignant gliomas but discontinued the research because of discouraging results. To deal with the penetration problem, researchers used neutrons with higher energy (up to 10 keV), called epithermal neutrons. These neutrons reached greater depths in tumors without causing significant damage to healthy normal tissue. Improved tumor targeting boron-10 compounds have been investigated.

A new therapeutic approach has been proposed by physicists that involves the use of gadolinium-157 instead of boron-10. Its neutron capture capacity is much greater than that of the boron-10 and it emits gamma rays in the tumor with an energy that is far greater than that of conventional X-rays [9]. Finally, after the emission of gamma rays, gadolinium becomes an isotope. We propose that the transport of gadolinium into the interior of the cells should be studied by a group of biophysicists.

There are no treatment/research centers for BNCT in Brazil. However, these centers exist in Argentina, the Czech Republic, Finland, Germany, Holland, Italy, Japan, Taiwan and the United States.

2.8   Which Therapy Must Be Used? Where?

Hadron therapy using carbon ions causes three times more damage to DNA strands than protons [7]. Therefore, carbon ions are superior to protons in killing cancer cells. As shown previously, proton therapy has a Bragg peak in the range of mm. The depth of the Bragg peak depends on the incident beam energy; the higher the energy, the deeper the Bragg peak. Broadening the Bragg peak to cover the entire volume of the tumor can be accomplished by employing a modulator that is placed in the output beam. Protons with an energy >160 MeV can reach tumors in the human body at depths of up to approximately 16 cm, which is sufficient for many types of cancer treatment, such as head and neck, ocular and prostate tumors. The Heidelberg Ion-Beam Therapy Center (HIT) at Heidelberg University Hospital, Heidelberg, Germany, has a heavy ion accelerator that provides ion beams consisting of carbon, oxygen 16 and others. The Frankfurt Institute for Advanced Studies (FIAS) in Germany has scientists from Brazil who are engaged in theoretical research regarding hadron therapy (such as Monte Carlo simulations). The neutron therapy unit at the Gershenson Radiation Oncology Center at the Karmanos Center/Wayne State University (KCC/WSU) is the most experienced center for prostate tumor therapy in the world. In South Africa, the NRF-iThemba Labs provide proton and neutron therapy.

Centers using protons/^{12}C:

- Heidelberg Ion-Beam Therapy Center (HIT) in Heidelberg, Germany.
- Centro Nazionale di Adroterapia Oncologica (CNAO) in Pavia, Italy.
- Heavy Ion Medical Accelerator (HIMAC) in Chiba, Japan.
- Hyogo Ion Beam Medical Center (HIBMC) in Hyogo, Japan.
- Gunma University Heavy Ion Medical Center (GHMC) in Gunma, Japan.
- Institute of Modern Physics, Chinese Academy of Sciences (IMP-CAS) in Lanzhou, China.

New proton/^{12}C therapy centers under construction:

- Proton/heavy ion facility in Marburg, Germany (Rhön Klinikum AG). Financial problems regarding this facility remain to be solved.
- Med-AUSTRON in Wiener Neustadt, Austria.
- Proton/heavy ion facility in Shanghai Hospital, Shanghai, China.

2.9 Loma Linda University Medical Center (USA) and Others

Below is a list of the diseases currently treated with protons [3, 11] (Figs. 2.6 and 2.7).

Fig. 2.6 Loma Linda University Medical Center. This was the first hospital-based proton therapy center. The first patient was treated in 2002

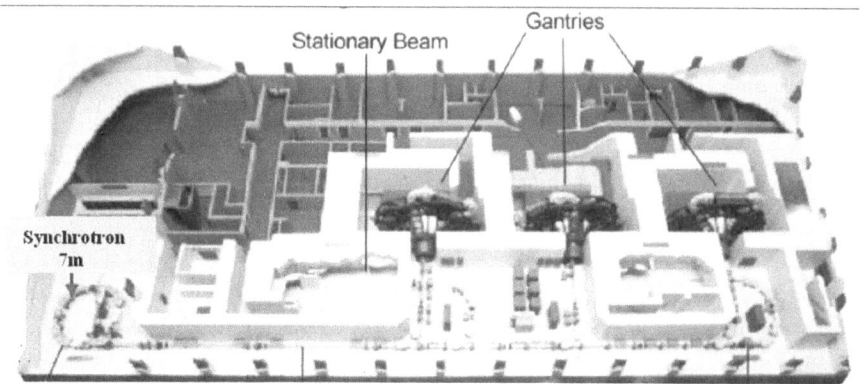

Fig. 2.7 The synchrotron (250 MeV) at Loma Linda University Medical Center used for proton therapy. There are three gantries, and in 2005, 160 sessions per day were possible

Brain and spinal cord

- Poor arteriovenous formations (AVMs)—treatment of defects of the circulatory system.
- Metastases isolated from the brain—high radiation doses reduce symptoms.
- Pituitary adenomas—fractioned radiation is given after radiosurgery.

Skull base

- Acoustic neuroma—benign tumors affecting hearing.
- Chondrosarcomas and chordomas—brain tumors that occur in the spinal cord or central nervous system.
- Meningiomas—tumors treated in a few sessions.

The eye

- Uveal melanoma—malignant tumors treated with protons to minimize the need for removal of the eye.

Head and neck

- Nasopharynx—local carcinoma treated with protons to reduce radiation side effects.
- Oropharyngeal cancer (locally advanced)—localization of radiation dose to minimize damage to healthy normal tissue.

Thorax and abdomen

- Recent lung cancer—local treatment minimizing damage to the normal lung.

Pelvis

- Prostate cancer—treatment with radiation at high doses. Local treatment aimed at high rates of survival and minimal collateral damage.

Tumors in children

- Brain tumors—highly individualized treatment.
- Ocular and orbital tumors—treatment to prevent injury to the lens and anterior chamber of the eye.
- Sarcomas at the base of the skull and spine—a variety of conditions are now treated in children.

Updated list of proton therapy centers within and outside the USA
It is a pleasure to acknowledge Christine Romero, head nurse at International Patient Care Services, Loma Linda University Medical Center, Loma Linda, CA, for the following information:
Proton Centers in the USA

1. Northeast Proton Center at Massachusetts General Hospital, Boston, MA.
   Website: http://www.massgeneral.org/cancer/about/providers/radiation/proton/index.asp
   Telephone: (617) 726 5130; toll-free (877) 726 5130; or (800) 388 4644

2. Midwest Proton Radiotherapy Institute at Indiana University, Bloomington, IN.
   Website: http://www.mpri.org/
   Telephone: (812) 349 5074; or toll-free (866) 487 6774
3. Proton Therapy Center, MD Anderson Cancer Center (Ranked # 1 as the best hospital for treating cancer in the USA [according to the US News and World Report (2011)]), Houston, TX.
   Website: http://www.mdanderson.org/care_centers/radiationonco/ptc/
   Telephone: (800) 392 1611
4. The University of Florida Proton Therapy Institute, Jacksonville, FL.
   Website: http://www.floridaproton.org/
   Telephone: (904) 588 1800; or toll-free (877) 686 6009
5. Procure Proton Therapy Center, Oklahoma City, OK.
   Website: http://procure.com/
   Telephone: (888) 847 2640
6. Proton Roberts Penn Center, Philadelphia, PA.
   Website: http://www.pennmedicine.org/perelman/proton/
   Telephone: (800) 789 7366

Proton centers outside of the USA

1. Rinecker Proton Therapy Center, Munich, Germany.
   Website:http://www.rptc.de/&inRussian;http://www.rptc.de/index.php?id=280&L=0
   Telephone: +49 (0) 89 660680
2. Svedberg Laboratory, Uppsala, Sweden.
   Website: http://www.tsl.uu.se/tsl_proton_narrow.html & http://www.tsl.uu.se/welcome.html
   Contact: Dr. Alexander Prokofiev; Telephone: +46 (0) 18 471 3850
3. Division of Radiation Medicine at the Paul Scherrer Institute (PSI), Switzerland.
   Website: http://radmed.web.psi.ch/
   Telephone: +41 (0) 56 310 3524
   Email: protonentherapie@psi.ch
4. Wanjie Proton Therapy Center, Zibo, China.
   Website: http://www.wanjiehospital.com/proton/index1.htm
   Telephone Number: +86 533 4650222; +86 533 4650000
   Fax number: +86 533 4650830

Since 1998, as reported by Ugo Amaldi [1], there has been a sharp increase in the use of hadron therapy, both in terms of the number of patients treated and the establishment of new care units. In Japan, the therapy has been developed furthest, especially in the case of carbon ions; however, in Europe, thanks to collaborations between nuclear research laboratories and cancer treatment hospitals, interest in hadron therapy has increased.

2.10 Heidelberg Ion-Beam Therapy Center (Germany) and Others

The HIT is located in Heidelberg, Germany. It is one of the world's best treatment centers for hadron therapy. A link to access the website is provided below. A full list of the diseases and possible treatments has not been provided in this Brief because the homepage is highly informative (in several languages) and provides a global view. It also gives access to information such as addresses, e-mails (contact), equipment, diseases and treatments, finance, leadership, and other information. There are also many photographs of the facility. Heidelberg is about 45 min by car or train from Frankfurt. Further information can be found at http://www.heidelberg-university-hospital.com/.

2.11 Procedures Used at HIT Before and During Treatment

The following procedures are used:

1. Before treatment begins, the tumor type and its dimensions are determined using CT and MRI scans.
2. The patient is held stable according to individual requirements. For example, in the case of brain tumors, a mask made of plastic material is used to keep the head still. For spinal tumors and other regions of the body there is a similar system.
3. The patient is placed on a high-technology operating table. An X-ray image is obtained to facilitate the alignment of the patient's position according to the coordinates obtained from the CT and MRI scans. Millimeter accuracy is required.
4. Radiation is delivered over a period of 5 min. The patient feels nothing. The tumor is scanned 20,000 times per second to check that everything is correct and the results are immediately visible on the monitor. This procedure is called the intensity-controlled raster scan method, and was developed by researchers at the GSI Helmholtz Center.
5. The total process described above takes on average 20 min. A treatment cycle takes approximately 15 days.
6. Several weeks after this treatment cycle, physicians check the progress of the therapy regarding tumor growth using CT and MRI scans.

2.12 Intensity-Controlled Raster Scan Method

We now examine the intensity-controlled raster scan method in detail (Fig. 2.8).

a

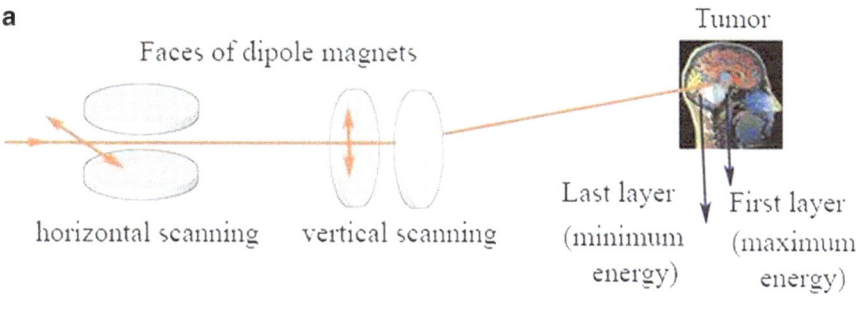

Faces of dipole magnets

Tumor

horizontal scanning   vertical scanning

Last layer (minimum energy)   First layer (maximum energy)

b

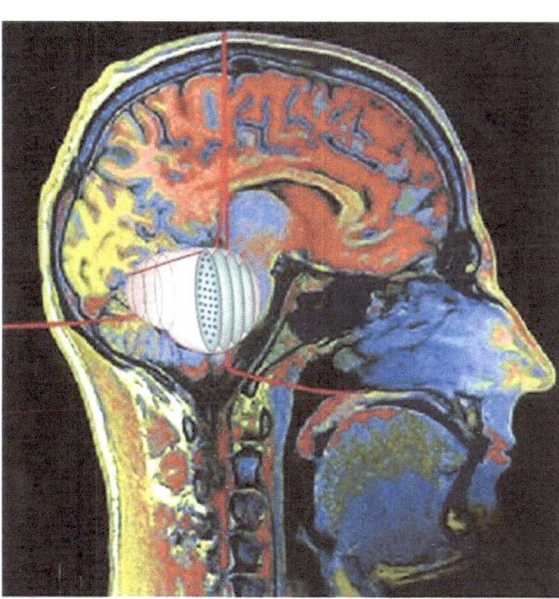

Fig. 2.8 (**a**) Schematic-picture of the intensity-controlled raster scan method (courtesy GSI [2.10]). (**b**) Details: layers and pixels

The tumor volume is subdivided into layers of equal depth. By varying the energy of the ions, the penetration depth of the beam can be adjusted. The individual layers are scanned line by line, in a similar manner to a TV screen. The lateral deflection of the beam is achieved by magnetic dipoles that can be controlled very quickly. To control the intensity, each row is divided into pixels. The beam remains at each pixel until a target dose is reached. As a result, this method allows a three-dimensional scanning of the tumor to be carried out, which is defined by the physician. This approach is a significant improvement over conventional treatment with photons and the application techniques of passive beams, which are employed in proton or ion beam therapy.

2.13   Modern Radiotherapy: Estimating the Risk
of Second Malignancies

Unfortunately, despite recent advances in radiotherapy, there is still the risk of cancer arising in a location that was previously free of disease caused by the treatment itself and not metastasis. There is always the possibility of developing cancer under these conditions because of secondary neutrons, which are inevitably produced in treatments involving particle beams [14]. Quantifying these risks requires a detailed knowledge of a range of parameters and a multidisciplinary team. In children, cancer is fortunately relatively rare. Radiotherapy is used to treat children with the following cancers: lymphoma, leukemia, brain tumors, sarcomas, Wilm's tumor, neuroblastoma, and liver cancer [12].

Traditional radiotherapy has been improved as a consequence of the development of IMRT, which has enabled improved targeting of conventional X-rays and a reduction in the radiation dose exposure of healthy normal tissue. It is the "state of the art" in photonics therapy. However, this technique is less effective than carbon ion therapy (as used at HIT). IMRT requires two to three times more monitoring units to deliver a specific radiation dose to the tumor target, when compared with conformal radiotherapy delivered in three dimensions (3D-CRT). Using IMRT instead of 3D-CRT increases the risk of developing secondary cancer by a factor of approximately 2. It is important to remember that particle therapy beams deposit most of their energy near the end of their tracks in the region of the Bragg peak. This peak is spread out to cover the entire tumor volume, and the dose beyond the tumor is lower than is the case in photon therapy. Currently, more than 20 centers worldwide are treating patients and more than 70,000 cancer patients have received particle therapy. Most have been treated using proton therapy, but the use of carbon ions is increasing. The neutrons produced during radiation therapy collide with protons in water and generate additional charged particles that can ionize the surrounding molecules. However, this problem can be addressed by using magnetically scanned beams rather than passively scattered beams. It is worth noting that the characteristics of cancer vary from organ to organ, and there is no evidence that the tumor dose–response curves are the same for different organs. Ionizing radiation is a carcinogen, as recognized by the WHO several years ago. In regions exposed to high doses, ionizing radiation directly kills cells in the field; however, the resulting tissue inflammation and DNA damage to cells in the normal tissue surrounding the tumor can promote cancer. We can say that in the radiotherapy of pediatric patients, the primary concern is low dose exposure to distal organs, while for adults, high radiation doses induce inflammation. To estimate the risk of developing cancer from protons, we must rely entirely on animal and in vitro cell experiments. Estimates of the RBE of neutrons are largely based on animal studies, although atomic bomb survivors have also been exposed to neutrons and some data are available. Finally, considerable uncertainty remains in predicting the late effects of heavy ions in humans. These ions are effective in inducing inflammation. In general, only the organs in the beam path are exposed to heavy ions, while the distal organs receive scattered neutrons and protons. In conclusion, there is good epidemiological evidence that radiation therapy can contribute to the long-term survival of children with cancer, but it also causes a high incidence of secondary malignancy among survivors. However, the data suggest that hadron therapy leads to a reduced risk of secondary malignancy as compared with conventional radiotherapy modalities that employ X-rays. We draw attention to the fact that when using heavy ions, the radiation dose to healthy normal tissues is very low. In addition, the production of neutrons by these ions is lower than is the case for protons, because fewer ions than protons are needed to achieve the same dose in the tumor target.

2.14   Elective Indications for Carbon Ion Therapy

- Chordoma/chondrosarcoma
- Malignant salivary gland tumors
- Malignant melanoma of the paranasal sinus
- Soft tissue sarcomas and bone tumors
- Early stage lung cancer
- Liver tumors
- Prostate carcinoma

For all these tumors, 90 % local control can be achieved, without severe complications.

References

1. Amaldi U, Bonomi R, Braccini S et al (2010) Accelerators for hadrontherapy: from Lawrence cyclotrons to linacs. Nucl Instrum Methods Phys Res A 620:563–577
2. Baccarelli I, Gianturco FA, Scifoni E, Solovýov AV, Surdutovich E (2010) Molecular level assessments of radiation biodamage. Eur Phys J D 60:1–10
3. Bert C, Durante M (2011) Motion in radiotherapy: particle therapy. Phys Med Biol 56:R113–R144
4. Bulanov SV, Esirkepov TZ, Khoroshkov VS, Kuznetsov AV, Pegoraro F (2012) Oncological hadrontherapy with laser ion accelerators. Phys Lett A 299:240–247
5. Durante M, Loeffler JS (2010) Charged particles in radiation oncology. Nat Rev Clin Oncol 7:37–43
6. Greiner AV (2009) Ion-induced electron production in tissue-like media and DNA damage mechanisms. Eur Phys J D 51:63–71
7. Grob KD, Laroudier CM, Peter I (2006) Ion-beam radiotherapy in the fight against cancer. GSI Brochure. www.gsi.de
8. Haettner E (2006) Experimental study on carbon ion fragmentation in water using GSI therapy beams. Master's Thesis, Kungliga Tekniska Hogskolan Stockholm. Available at EXFOR: http://www-ds.iaea.org/exfor/exfor.htm
9. Hosmane NS, Maguire JA, Zhu Y, Takagaki M (2012) Boron and gadolinium neutron capture therapy for cancer treatment. World Scientific Publishing, Singapore

10. Krämer M, Durante M (2010) Ion beam transport calculations and treatment plans in particle therapy. Eur Phys J D 60:195–202
11. Mishima Y (ed) (1996) Cancer neutron capture therapy. Springer, New York
12. Newhauser WD, Durante M (2011) Assessing the risk of second malignancies after modern radiotherapy. Nat Rev Cancer 11:438–448
13. Surdutovich E, Solovýov AV (2012) Double strand breaks in DNA resulting from double ionization events. Eur Phys J D 66:206
14. Timlin C, Jones B (2010) Proton and charged particle radiotherapy. Br J Radiol 83:87

Chapter 3
Physics

3.1 Particle and Ion Accelerators

3.1.1 Brief History

The invention of the cathode ray tube (also known as the cinescope) by Karl Ferdinand Braun in the period 1895–1897 was an important step in the development of particle and ion accelerators. The cathode ray tube can be considered the first particle accelerator that used a high electric potential difference. Its development led to the discovery of the electron by Joseph J. Thomson in 1897. In 1911, Ernest Rutherford (the father of nuclear physics) and colleagues proposed the idea that the atomic nucleus was surrounded by electrons, which was a different concept to the model proposed by Thomson.

There are three types of accelerator in regular use, namely the potential difference, linear and circular types [5]. To build an accelerator that could make a potential difference, the Cockcroft–Walton voltage multiplier was installed at Fermilab in the USA. It reached a maximum energy of 750 keV, and in 1932 the first artificial disintegration of the type $^1_1H + ^7_3Li \rightarrow 2.^4_2He$ was obtained using this accelerator.

The best-known accelerator regarding potential difference was constructed by an American engineer of Dutch descent named Robert Jemison Van de Graaf. This 40-m-tall accelerator achieved a maximum energy of 25 MeV (Fig. 3.1).

M. d'Ávila Nunes, *Hadron Therapy Physics and Simulations*, SpringerBriefs in Physics, DOI 10.1007/978-1-4614-8899-6_3, © Marcos d'Ávila Nunes 2014

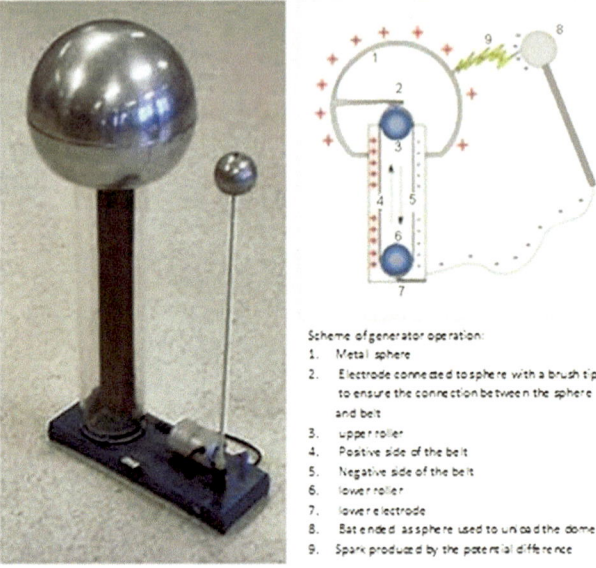

Scheme of generator operation:
1.   Metal sphere
2.   Electrode connected to sphere with a brush tip
     to ensure the connection between the sphere
     and belt
3.   upper roller
4.   Positive side of the belt
5.   Negative side of the belt
6.   lower roller
7.   lower electrode
8.   Bat ended as sphere used to unload the dome
9.   Spark produced by the potential difference

Fig. 3.1 Van de Graaf generator. Wikimedia Commons

Following this development, an accelerator with a maximum energy of 500 MeV
was built. This was more compact and was used at the Oak Ridge National Laboratory
to produce uranium 235, which was used in the atomic bomb. It was called the
Tandem Van de Graaf. In 1974, the Institute of Physics at the University of São Paulo
in Brazil constructed a tandem accelerator, the first one in Latin America (Fig. 3.2).

Fig. 3.2 Photograph of a tandem Van de Graaf accelerator (1970). It consisted of two electrostatic
accelerators connected in series. Each was 24 m long and generated 15,000,000 V. Wikimedia
Commons

The linear accelerator (linac) was proposed in 1924 by physicist Gustav Ising, and was built 4 years later by the physicist Rolf Wideröe. It was a system that allowed particle acceleration in the spaces between consecutive cavities that were interconnected by an oscillating voltage source. However, to reach very high energies, the accelerators needed to be built to extend over a distance of >3 miles (Fig. 3.3).

Fig. 3.3   Conseil Européen pour la Recherche Nucléaire (CERN) LINAC 1. Wikimedia Commons

Circular accelerators were called betatrons, cyclotrons, synchro-cyclotrons, synchrotrons, tevatrons and microtrons. However, cyclotrons and synchrotrons have prevailed as the accelerators of choice. These two types of accelerators are discussed in detail below.

3.2   The Cyclotron

The cyclotron was invented in 1932 by Ernest Lawrence. It accelerated protons with a fixed frequency of up to 1.25 MeV, allowing nuclear transmutation [3, 5]. Lawrence received the Nobel Prize 7 years later. The University of Berkeley recognized the potential of this new machine and built a 5-m-long cyclotron that accelerated protons to an energy of 20 MeV.

Figure 3.4 shows two current cyclotrons and a schematic drawing in which the magnetic field imposes a circular path on the particles. The oscillating electric field (RF) is responsible for particle acceleration; the final trajectory is a spiral.

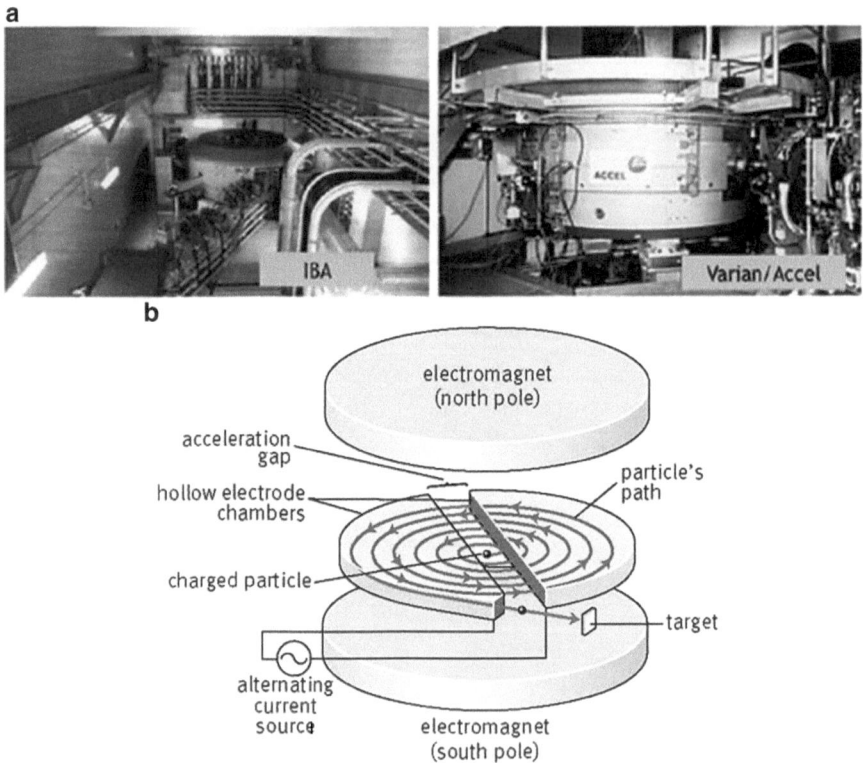

Fig. 3.4 (**a**) Commercial cyclotrons (IBA variants and the Varian/Accelerator). (**b**) Schematic drawing of a cyclotron

3.2.1 Motion of Particles: Equations

The force acting on a particle, with velocity v in a magnetic field B, has the following characteristics. The direction is perpendicular to the plane (v, B); this can be represented by the left hand rule where the thumb, index and middle fingers are mutually perpendicular (90°). The thumb indicates the direction of the force, the index finger the vector magnetic field and the middle finger the speed. The magnitude of the magnetic force that acts on the particle is given by $f = q.v.B\sen\phi$, where f is the magnetic force, q is the particle charge in Coulombs, and ϕ is the angle between the vectors B and v (this angle can vary from 0 to 180°). This is explained in more detail in Fig. 3.5 [Nunes, M.A. (2013) Large Hadron Collider—New Era of Discoveries, in press].

Fig. 3.5 Plans and layout of
the magnetic field vectors,
velocity and magnetic force
acting on a proton

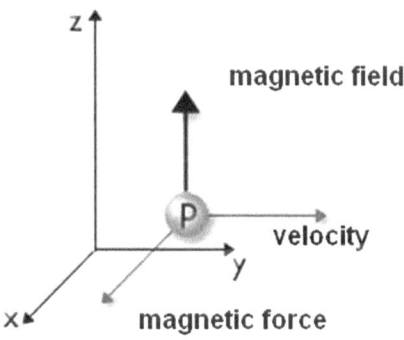

   Furthermore, when observing the particles from a proton–proton collision, for
example, it can be concluded that because the magnetic field can bend the path of a
particle, particles of lower energy bend more and higher energy particles bend less.
Because the magnetic force depends on the active particle charge, the trajectory can
curve in either direction. If v is perpendicular to B, the equality decreases to
F=q.v.B. However, Newton's equation gives the expression for the force F=mass
(m)×acceleration (a), which means that to find the necessary replacements, the
following equations are used for the speed and radius:

$$v = q.B.r \, / \, m \text{ and } r = m.v \, / \, q.B$$

   Because we know v, we can calculate the acceleration as: $a = v^2/r$
   Substituting the value of v in the equation, the acceleration is obtained. Because
the radiated power is given by P $\alpha(q^2.a^2/c^2)$, that is P $\alpha(q.B/m)^4.r^2$, it can be concluded that the smaller the mass, the greater the radiated power which causes the
withdrawal of the beam particle. Thus, the electron radiates more power than the
proton.

3.2.2   Calculating the Frequency of the Cyclotron

Calculation of the frequency of the cyclotron is very straightforward. The total turn
is equal to $2\pi.r$ in a path of radius r. If t is defined as the time spent in the half turn:
$v = \pi.r/t$, then $t = \pi.r/v$; however, because $v = q.B.r/m$, then, $t = \pi.m/q.B$. Thus, the
time spent on the course is the same for all orbits, independent of the radius. Because
the period of one complete turn (T) is twice that spent in the half turn T=2.t, then
$T = 2\pi m/qB$. Because the frequency (ν) is the inverse of the period, we have $\nu = 1/T$,
and thus $\nu = qB/2\pi m$.
   The angular frequency becomes $\omega = 2\pi\nu$, then $\omega = qB/m$.
   This is the frequency value obtained from the RF source to produce the acceleration of a charged particle q, and mass m, which are subjected to the magnetic field

B. It can be concluded that the cyclotron frequency is directly proportional to B and inversely proportional to the ratio m/q. Thus the particle with lowest m/q ratio produces a spiral with more full turns (higher frequency), provided that the field remains constant.

A video regarding the cyclotron can be viewed using the link below: http://www.youtube.com/watch?v=cNnNM2ZqIsc&feature=related

If a cyclotron (200 MeV) were as small and inexpensive as one of the 5–20 MeV linacs used in conventional radiotherapy, then more than 90 % of patients could be treated with a proton beam. The accelerators used today are large and expensive, costing around 20 M euros for a proton accelerator and 40 M euros for a carbon ion beam facility. The installation of gantries would add another 10–12 M euros to the cost. The gantry used at the Heidelberg Ion-Beam Therapy Center (HIT) weighs 670 tons and consumes 400 kW of power. Considering the large hadron collider (LHC) complex as a whole, the power consumption would be approximately 120 million watts of electrical power at peak demand. The stored energy is 11 billion joules [Nunes, M.A. (2013) Large Hadron Collider—New Era of Discoveries, in press]. In the future, it is possible that gantries will be built using superconducting magnets. The current situation regarding size and costs is expected to change in the future. A Belgian company, IBA, already offers a superconducting cyclotron with a 6 m diameter, which accelerates carbon ions up to 400 MeV/u.

TERA introduced and developed a new type of accelerator, the cyclinac [2], to accelerate protons and carbon ions, with a time required to vary the energy of the beam of only 1 ms, compared with the 20–50 ms needed by a cyclotron and a the1 s needed by a synchrotron.

3.3 The Proton Synchrotron

Conceptually, the principle concept of the synchrotron was published in a Russian scientific journal by Vladimir Veksler. He established the journal entitled "Nuclear Physics" (*Yadernaya Fizika*) and became its first editor-in-chief. The first synchrotron was built by Edwin McMillan in 1945, and the first proton synchrotron was designed by Sir Marcus Oliphant and built in 1952. In particle physics, the synchrotron is a cyclic particle accelerator in which the electric field is responsible for the acceleration of particles, and the magnetic field is responsible for the change in direction of the particles; both fields are synchronized with the particle beam.

An interesting comparison was proposed by Don Lincoln in his book "The Quantum Frontier" [9], which facilitates an understanding of the functioning of a proton synchrotron. The principle that governs this accelerator is the same as that which governs a tetherball. A tetherball is a ball attached to one end of a rope. The other end is attached to the top of a tall pole, which is anchored deep in the ground.

A person hits the ball and the rope ensures that the ball travels in a circular path. Once the ball makes a full circle, it is hit again. The ball goes faster and makes another circuit. If the rope is attached to the top of the pole, it does not wrap itself around the pole; in principle, the ball can be made to travel very rapidly by synchronizing both its orbit and the person hitting it (hence the derivation of the name synchrotron). In a proton synchrotron, the electric field "hits" the proton and accelerates it. However, the counterpart of the rope in the tetherball analogy is not provided by electric fields, but rather by magnetic fields. Particles are accelerated by an electric field over a short distance, and are then guided by magnetic fields in a circular path back to the electric field region for another round of acceleration. The synchrotron was based on the cyclotron with a time-dependent magnetic guide, which was synchronized to a particle beam with increasing kinetic energy. The difference between the cyclotron and synchrotron is that the latter uses the principle of phase stability, maintaining the synchronism between the applied electric field and the frequency of revolution of the particle. Beam focusing and acceleration can be separated into different components using the curvature of the synchrotron beam. Thus, radio frequency cavities are used for acceleration, the magnetic dipoles for particle deflection, and quadrupole/sextupole magnets for focusing the beam particles. The magnetic field maintains the orbit instead of accelerating the particles, and hence the magnetic field lines are only necessary in the region defined by the orbit.

The synchrotron facility consists of the following components (Fig. 3.6a): (1) The ion source accelerator. This is where ion beams composed of positively charged atoms are produced. For protons, hydrogen gas is used. For carbon ions, dioxide is used; (2) A two-stage linear accelerator. Ions are accelerated in structures at high frequency up to 10 % of the speed of light; (3) The synchrotron. Six 60° magnets bend the ions into a circular path. After a million orbits, ions are accelerated to 75 % of the speed of light; (4) The treatment room beam lines. Magnets guide and focus the beam of ions in vacuum tubes; (5) The treatment room. The beam enters the treatment room through a window. The patient is positioned on a treatment table that is adjusted accurately by a computer-controlled robot; (6) Position control. Using a digital X-ray machine, images are obtained before irradiation. The computer software compares the images obtained with those used in treatment planning and precisely adjusts the position of the patient; (7) The gantry. The rotation system enables the beam to be directed toward the patient at an optimized angle. The gantry weighs about 670 tons, 600 tons of which can be rotated with submillimeter accuracy; (8) The treatment room in the gantry. This is where the beam exits the gantry (beam line). Two rotation systems and digital X-rays are used to optimize the position of the patient guided by the images taken before irradiation.

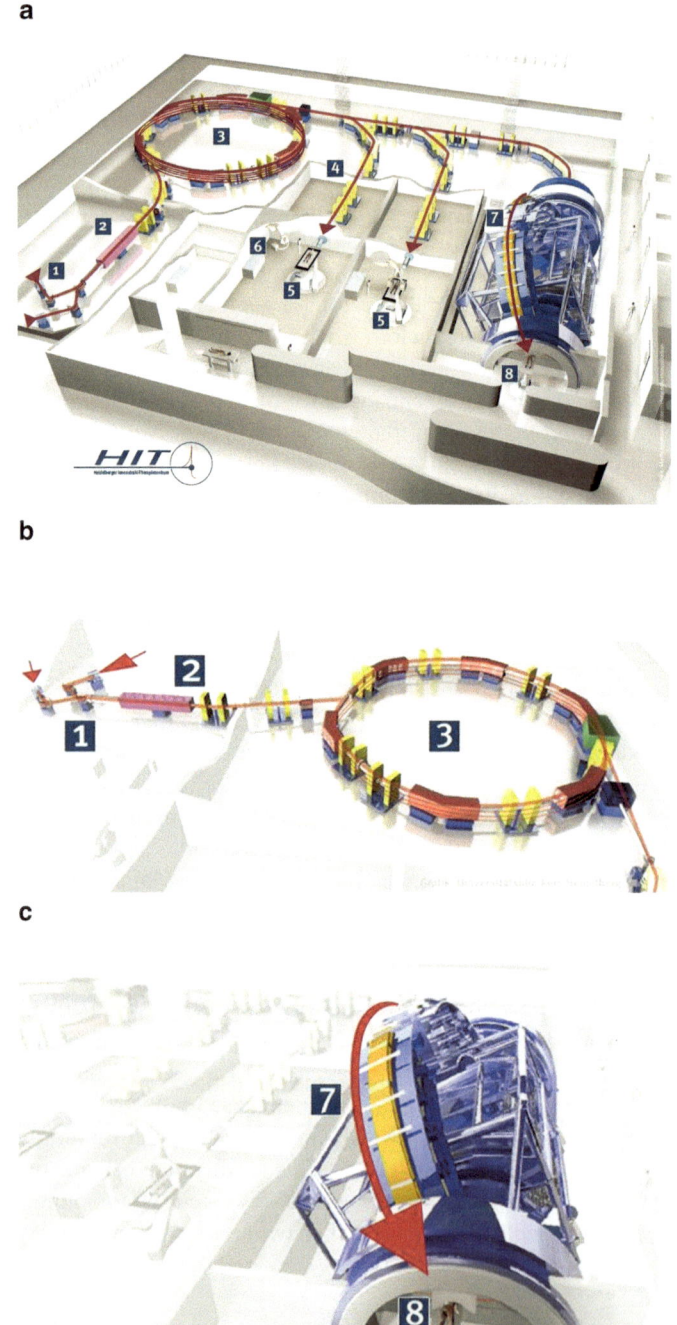

Fig. 3.6 (**a**) Synchrotron at the HIT in Heidelberg. *Source*: Heidelberg University Hospital. (**b**) Schematic of the synchrotron at the HIT. (**c**) Schematic drawing of the HIT gantry. Courtesy Annette Tuffs, Head of Corporate Communications/Press Office

The combination of magnetic field "guides", time-dependency and the principle of strong focus enables the design and operation of modern large-scale accelerators and colliders and even synchrotron light sources, such as at the Brazilian Synchrotron Light Laboratory in Campinas. The power limit could be increased by using superconducting magnets, which are not limited by magnetic saturation. Electron and positron accelerators may be limited by the emission of synchrotron radiation, resulting in a partial loss of kinetic energy of the particle beam. Therefore, the energy of electron and positron accelerators is limited by the loss of this radiation, which does not happen with proton or ion accelerators. The energy of these accelerators is limited by the strength of the magnets and the cost.

In the synchrotron, particle injection is pre-accelerated using a linac, microtron or even another synchrotron, because synchrotrons are unable to accelerate particles from zero kinetic energy.

The tevatron at Fermilab was the largest collider in the world in 2008. It accelerated protons and antiprotons to 1 TeV and then collided them. The LHC has seven times this energy; accordingly, the proton–proton collisions occur at about 14 TeV. The LHC also accelerates heavy ions (such as lead) up to an energy of 1.15 PeV [7].

Possible applications of the synchrotron

Possible applications of the synchrotron include the following:

1. In biological sciences, the study of protein and crystallography.
2. New drug development and research.
3. "Firing" of drawings from a computer chip onto a metal sheet.
4. Investigating the composition of chemical compounds.
5. Evaluation of the reactions of living cells to drugs.
6. Crystallography of inorganic material and microanalysis.
7. Fluorescence studies.
8. Semiconductor material analysis and structural studies.
9. Geological material analysis.
10. Medical imaging.
11. Ion therapy (mainly carbon ions) for the treatment of cancer and resistant tumors.

3.4   The Synchrotron Light Source

Synchrotron radiation is produced when high-energy particles are in fast motion, including electrons forced to travel in a curved path by a magnetic field. Synchrotron radiation is proportional to the fourth power of the particle velocity and inversely proportional to the square of the radius of the path (trajectory). Synchrotron radiation is also observed in astronomical objects such as pulsars (plerions). A typical example of this is the Crab Nebula. In this case, the synchrotron radiation is produced by electrons trapped in strong magnetic fields around the pulsar (Fig. 3.7).

Fig. 3.7 The *blue* color in the central region of the nebula (Crab Nebula) is caused by radiation similar to that generated by a synchrotron. Wikimedia Commons

3.4.1 Water Channel Research Using a Synchrotron Light Source

The biochemists Peter Agre and Roderick MacKinnon used synchrotron light sources as part of their research, for which they were awarded the Nobel Prize in Chemistry in 2003. They studied the processes involved in water flow across cell membranes and cell communication. This work has helped in understanding the molecular pathways of disease (Fig. 3.8).

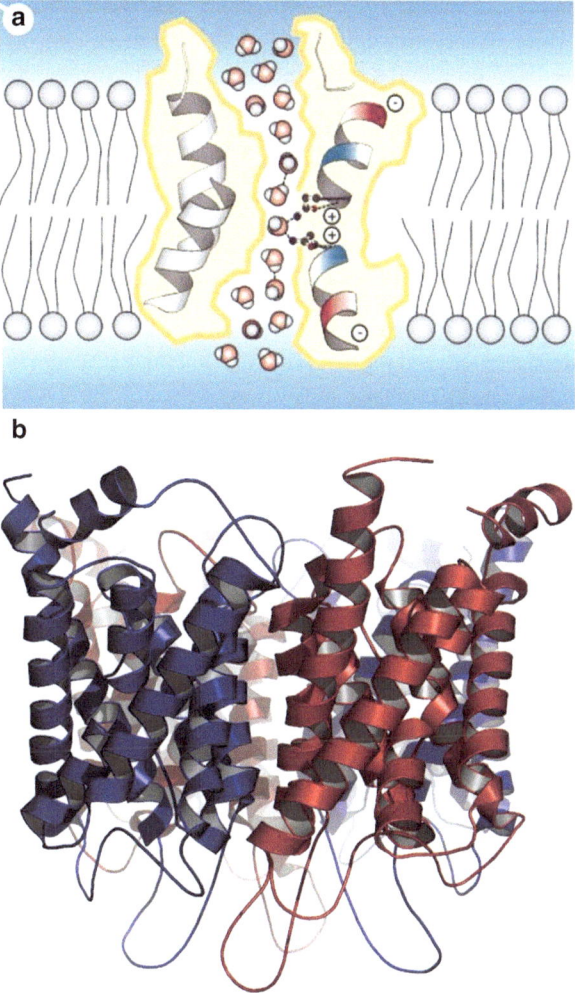

Fig. 3.8 (**a**) Aquaporins (integral membrane proteins that regulate the flow of water) embedded in the cell membrane. Courtesy Yoshinori Fujiyoshi. (**b**) Crystallographic structure of aquaporins. Wikimedia Commons

Aquaporins are integral proteins that form pores in the membranes of biological cells that regulate the flow of water. Genetic defects involving aquaporin genes have been associated with several human diseases. The Nobel Prize in Chemistry in 2003 was awarded to Peter Agre for the discovery of aquaporins, and to Roderick MacKinnon for his work on the structure and mechanisms of potassium channels. Aquaporins selectively conduct water molecules into and out of cells, which prevents the passage of ions and other solutes; they act as water channels. Interestingly, there are also aquaglyceroporins which carry small solutes such as glycerol, CO_2, ammonia and urea through the membrane, depending on pore size. The pores are completely water-impermeable to charged species such as protons, which is a critical property for the preservation of the membrane's electrochemical potential. The water molecules pass through the membrane via the pore channel in single file. The presence of water channels increases the membrane's permeability to water and is fundamental for the water transport system [10].

As mentioned earlier, synchrotrons are employed in the biological sciences to study proteins and crystallography. Thus, they serve as a useful tool for the structural study of integral proteins such as aquaporins. Certainly, the synchrotrons used are similar to the type that we have in Campinas, Brazil (Fig. 3.9); in other words, the synchrotron light sources are similar in terms of the output scans from infrared to X-rays. Proteins that are extracted from biological systems are purified, crystallized, cryotreated and analyzed under X-rays, so that their structure is revealed. This is the approach that Peter Agre used to study aquaporins with a synchrotron.

Fig. 3.9 The synchrotron light source at Campinas, SP, Brazil. Wikimedia Commons

The synchrotron in Campinas accelerates electrons and not protons. In the basement of the complex is an electron gun that shoots out the electrons. These are accelerated by an electric field (radio frequency cavities) and are guided by magnetic fields up to the top floor where the larger equipment is situated; they gain acceleration as they travel and eventually reach speeds close to the speed of light. It has been shown in the equations for particle motion that the smaller the mass of the particle, the higher the radiation produced by acceleration; therefore, electrons radiate more power than protons. Employing quadrupole or sextupole magnets, it is possible to converge the electrons and focus them to hit a target (specified in advance). The radiation resulting from their acceleration, scanning from infrared to X-rays, can be used to study protein crystals, and for numerous other applications as listed above. The synchrotron at Campinas works with radiation derived from the acceleration of electrons to relativistic velocities; however, for hadron therapy, a synchrotron designed for accelerating protons or carbon ions (and possibly other particles) is required.

A good exercise to test our knowledge would be to evaluate whether it would be possible to convert a synchrotron light source into a proton synchrotron. What modifications would be needed? For example, to start with, the electron gun would have to be replaced by a duoplasmatron. A key question is whether such a general transformation would be worthwhile economically. This is discussed in greater depth in ref. [13].

The following points need to be considered:

1. The electron gun would have to be replaced by a duoplasmatron to produce protons.
2. The linac would have to be prepared so that it could accelerate protons and not electrons.
3. A "booster" for the protons would have to be prepared.
4. The superconducting electromagnets would have to undergo reconfiguration.
5. The protons emit a very small amount of radiation as compared with the electrons, for reasons explained previously. They could only be used for collisions and not to generate radiation.
6. The quadrupole or sextupole electromagnets necessary to focus the protons would have to undergo a new configuration.
7. All equipment for the use of radiation would have to be withdrawn.
8. The use of the synchrotron for hadron therapy would require the construction of three rooms: one for the patient, with the gantry allowing access to the tumor in any position; one for control; and one for research. The ideal setup would be three rooms for the gantry, allowing several treatments at the same time. A system like that is installed at Loma Linda University Medical Center in the USA.

In addition to these eight items, we propose to:

1. Use the intensity-controlled raster-scan method.
2. Use a system for gating, to work with moving targets such as lung tumors and hepatocellular injury; this will be discussed further in connection with the analysis of target motion in radiotherapy.
3. Hire oncologists and specialists prepared at CERN for hadron therapy.
4. Provide national and international public information about the proton synchrotron via TV and the internet, for patient care and research.
5. Provide courses for doctors, physicists and interested lay persons.
6. Establish maintenance contracts with companies such as IBA, PSI and Siemens.
7. Organize international exchanges with: CERN, Switzerland; CNAO, Italy; LLUMC, USA; HIT, Germany; and other centers to enable experts to exchange views at conferences and meetings.

Conclusion: Clearly, it would be a disadvantage to introduce changes in a synchrotron light source to convert it into a proton synchrotron.

3.5   The Cyclinac

The name cyclinac is a combination of cyclotron and linac. The cyclinac consists of a linac with a high frequency and a fast cycle that increases the energy of the particles previously accelerated by a cyclotron. The cyclinac can easily accelerate currents of the order of 2 nA, which are required for proton beam therapy (carbon ions), producing optimized ion beams for the irradiation of solid tumors using the most modern techniques [2]. The accelerators used for proton therapy are cyclotrons that are 4–5 m in diameter, and synchrotrons of 6–8 m diameter. For carbon ion therapy, only synchrotrons of 20–25 m diameter are employed. Recently, large superconducting cyclotrons have been built for carbon ion acceleration.

In summary, cyclinacs are excellent accelerators for hadron therapy [4] because of the following characteristics: they work with frequencies of 300 Hz, allowing efficient tumor mapping; they have a low power consumption of 800 W (reducing costs); they rapidly modulate the active energy (1 ms) that is essential for studying organs in motion; and finally, they have steep acceleration gradients and have a reduced size (a cyclotron weighs 190 tons and a linac is 24 m long). The compact cyclotron used for the first particle acceleration to 120 MeV/u is smaller than the more widely used cyclotron for proton therapy (e.g., IBA C235). Figure 3.10 shows the cyclinac (CABOTO) machine proposed by the TERA Foundation.

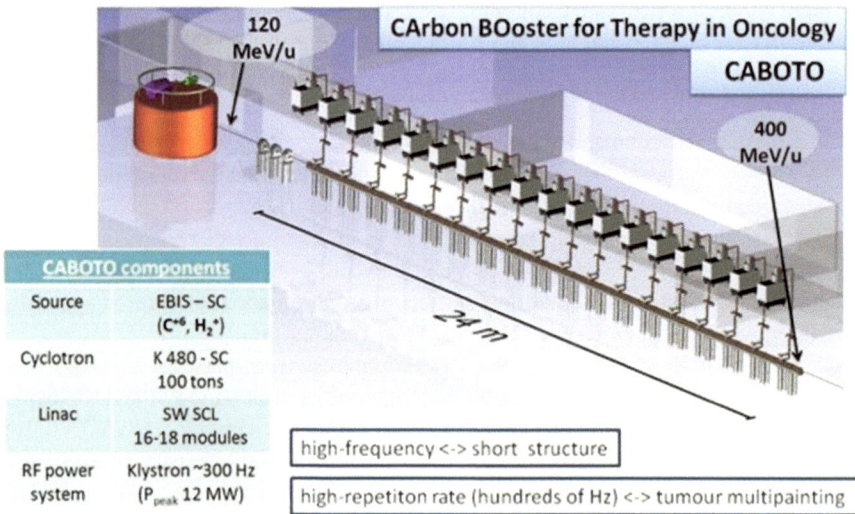

Fig. 3.10   TERA proposal: The cyclinac [2]. Courtesy Ugo Amaldi

3.6   How to Obtain Particles (Protons, Neutrons) and Heavy Ions for Hadron Therapy

Below we explain how to prepare protons, neutrons and heavy ions for use in hadron therapy [Nunes, M.A. (2013) Large Hadron Collider—New Era of Discoveries, in press].

3.6.1   How Are Protons Obtained?

Protons are produced by applying an arc discharge of hydrogen gas into a source called a duoplasmatron. The electron is released from the hydrogen atom, leaving the positive nucleus, a proton, floating freely in the resulting plasma. By applying a strong electric field, the protons are extracted from the plasma surface and are sent on their way as a stream of positive particles. Currents of up to 300 mA can be obtained.

The protons interact with matter in three distinct ways:

1. They slow down through collisions with atomic electrons and finally stop.
2. They are deflected by collisions with atomic nuclei causing scattering.
3. Collisions with a nucleus yield secondary particles in motion. This is called nuclear interaction.

The first two conditions occur via electromagnetic interaction between the charge of the proton and the charge of the electrons or the atomic nucleus. There are mathematical theories for the first two conditions. Nuclear interactions are known to be infrequent and function according to a set of models. Even though computer programs are used to solve these problems, which are accurate to within seconds, predicting the dose for a patient is very complex and time-consuming.

3.6.2   How Are Neutrons Obtained?

Neutrons can be obtained by accelerating deuterons with an energy of 48.5 MeV onto a beryllium target. The deuterons are accelerated using a superconducting cyclotron. Generally, neutrons can be obtained by accelerating protons or deuterium and colliding them with a beryllium or lithium target, provoking reactions of the type ^9Be (p, n) ^9B, ^7Li (p, n) ^7Be, ^3H (^2H, n), and ^4He (where p = proton, ^2H = deuterium, n = neutron, Be = beryllium, B = boron and ^4He = helium-4).

James Chadwick discovered the neutron in 1932 using alpha particles (from radioactive polonium), with which he bombarded a blade of beryllium. He noted that uncharged particles left the beryllium bulwark. He placed in its path a paraffin bulwark from which the protons were discharged after bombardment by particles without being charged. The neutron was discovered! Its mass was determined in a way that was very similar to the proton, because the impact removed protons from paraffin. The discovery of the neutron triggered a considerable increase in knowledge regarding nuclear structure. Additionally, in 1932, Werner Heisenberg realized that the nuclei of atoms were composed of protons and neutrons. He described the quantum mechanics involved and received the Nobel Prize in Physics in 1932. James Chadwick also received the Nobel Prize in 1935 for his discovery of the neutron. After the discovery of the neutron, Robert Stone began clinical trials with fast neutrons (radiation therapy) at the Lawrence Laboratory in Berkeley, CA, USA.

3.6.3   How Are Heavy Ions Obtained?

Heavy ions are atomic nuclei that have lost their electrons and are heavier than protons (hydrogen nuclei). A variety of ions are used, such as helium, carbon and oxygen nuclei. Heavy ions are three times more effective than protons and helium ions. In the human body, heavy ions can be targeted with millimeter precision, and are therefore superior to protons in the treatment of certain tumors. As is well known, ions are charged atoms. Thus, to obtain ions, atoms must necessarily lose their negatively charged electrons. For this purpose, carbon dioxide gas flowing within an ionic chamber is used. Free electrons in the gas are accelerated using magnetic fields and microwaves. Traveling through the ionic chamber, the electrons impact the molecules of carbon dioxide. After a collision, the molecules dissociate, and four of the six electrons in the carbon atom are separated. Electric fields are then employed to extract the carbon ions from the chamber. A special magnet transports them in a vacuum in a steady flow. This flow is converted into a pulsating flow with a frequency of 217 million pulses per second. The beam is collimated and the ions are accelerated. Subsequently, electromagnetic fields accelerate the ions to more than 10 % of the speed of light. Leaving the accelerator through a sheet of carbon, the carbon atoms lose their last two electrons, so that only nuclei with six positive charges remain.

3.7   Comparison of the Cyclotron, Synchrotron, Linac and Cyclinac

Currently, hadron therapy centers are only using circular accelerators such as cyclotrons and synchrotrons. Proton therapy centers employ synchrotrons and cyclotrons; however, only synchrotrons are employed for carbon ion therapy because of the higher energy required and the magnetic rigidity. The cyclotron works with energy in the range of 230–250 MeV, while synchrotrons work with a higher energy in the range 400–450 MeV. The beam is always present in cyclotrons, and the energy is not electronically adjusted. Moreover, energy can be varied in steps of 10 ms, with pulses separated by 10–20 ns; this makes cyclotrons excellent for working with organs in motion. This represents a major difference to the synchrotron. Although a synchrotron's power can be electronically adjusted, it requires 1 s to vary its beam energy, which is required to decrease the magnetic field and to accelerate the particles to the desired energy. In addition, the beam is not always present as is the case in the cyclotron. Thus, because the frequency of the synchrotron beam is similar to that of the respiratory cycle, it can be concluded that this represents a drawback for the irradiation of moving organs [8]. It is therefore unsuitable for use with the respiratory gating technique.

Linacs allow the beam energy to be varied in steps of 1–2 ms with electronically adjusted energy and a constant beam. In cyclinacs, energy can be adjusted between the output value of a cyclotron and the maximum possible output value of the linac, which is adequate for respiratory gating. Furthermore, cyclinacs work with high energy, and it is therefore possible to use them for carbon ion therapy.

Taking Europe as an example, out of a total population of 10,000,000, about 20,000 cancer patients per year are treated using conventional radiation therapy with X-rays, delivered in approximately 30 sessions on average per patient. Radiation therapy using protons (12 %) is received by only 2,400 patients per year, and a total of 24 sessions are needed on average per patient. About 600 patients per year are treated with carbon ions (3 %), and a total of 12 sessions are needed on average per patient. Why this difference? Linacs generating electrons with an energy of 10 MeV, which are used in conventional radiotherapy, are smaller and less expensive than cyclotrons or synchrotrons. If this were not the case, it is possible that 90 % of patients would be treated with proton beams. Circular accelerators are large and expensive and in general need 3–5 treatments rooms. A treatment using proton therapy is approximately 2.5 times more expensive than a treatment using IMRT with X-rays.

3.8   Use of Lasers in Hadron Therapy

Improvements in heavy ion therapy can be achieved using less expensive acceleration technologies. For potential use in therapy, the laser pulses need to accelerate protons to energies of ≤ 150 MeV and carbon ions to an energy of 350 MeV. Working with proton beams and focusing them is costly and difficult. Eccentric and isocentric equipment is used to transport proton beams from the final section of the accelerator to the target tumor. These structures are made of heavy magnets used to deflect the beam; these weigh 100–200 tons and have a diameter of 4–10 m. All this equipment can generate costs of up to 150,000,000 euros. Petawatt class lasers are not necessarily much smaller than the conventional accelerators used for hadron acceleration. The targets used to generate protons are only a few cm in size. Therefore the target is positioned close to the patient, using small mirrors to transport the laser instead of heavy and expensive magnets. This makes the equipment much lighter, smaller and less expensive. The laser beam can be sent to different treatment rooms using mirrors. In addition, the safety system for the laser is simple and inexpensive and damage to the eyes of the doctor and the patient is prevented. An additional advantage is that the laser accelerator does not require radioprotection with thick concrete walls. The repetition rate of lasers will soon be increased to kHz, and petawatt lasers with diode-pumps (used in Germany) and fiber lasers will have power and repetition rates >100 Hz.

Fourkal et al. [6] showed in 2002 that under conditions of optimal interaction, protons can be accelerated to relativistic energies of 300 MeV using petawatt lasers. The protons are accelerated by means of the Coulomb force, which arises from charge separation induced by high-intensity lasers. The proton energy and phase spatial distribution obtained from the particle simulations in cells are used to calculate the dose distribution using the GEANT Monte Carlo simulation code. Because of the wide range of energy and the angular spectrum of protons, a compact particle selection and beam collimation is necessary to generate small poly-energetic beams of protons for modulated-intensity proton therapy.

Intense and collimated proton beams produced by a high-intensity laser pulse interacting with a plasma were first developed for proton therapy of malignancies by Bulanov et al. [4]. The fast proton beam was produced by directing a laser at the target, which generated accelerated proton beams of high quality. A simple comparison between the traditional accelerators and laser accelerators shows the superior qualities of the laser.

3.8.1   How Are Protons Accelerated with a Laser?

It is possible to accelerate protons by means of a violent acceleration of electrons in the laser field that draws protons behind them on the posterior surface of the target (Fig. 3.11). This creates a continuous proton spectrum. Computations have shown that by using two appropriately shaped targets, a scattering energy of 3 % can be achieved. In Fig. 3.11, a powerful laser pulse is shown acting violently on a target constituted by a thin blade doped with hydrogen atoms [12]. The laser accelerates electrons off the posterior region of the target, creating an electric field that favors the output of protons from the target. In the future, it is hoped that laser pulses with intensities in the range 10^{18}–10^{20} W/cm^2 and pulse durations of 30–50 fs will be possible [1]; this will allow a facility for treatment with protons (single facility) to be constructed based on illumination of a thin target.

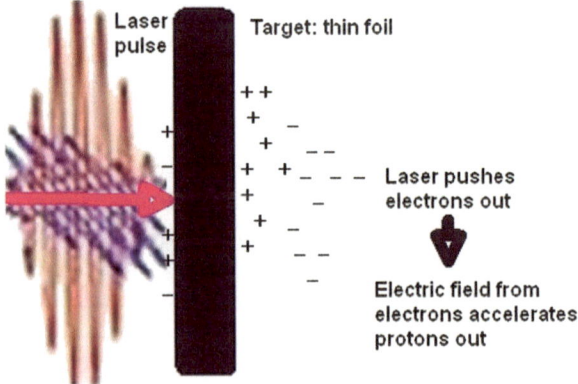

Fig. 3.11 Schematic drawing of a powerful laser blast onto a thin solid blade doped with hydrogen

Some companies are working to reduce the size and cost of high-power lasers, and there are several projects focused on improving beam quality. It is possible that in the next decade, the cyclotron will become obsolete and will be replaced by more compact laser systems. Many years of dedicated research are needed to achieve this goal; for now, it is not considered to be economically advantageous.

3.9   The Dielectric Wall Accelerator

Dielectric wall accelerators (DWA) are a type of induction accelerator. A traveling high-gradient field is created by switching high voltages on electrodes that are sandwiched between high-gradient insulators. This is the operating principle of the DWA. The accelerator tube is made of fused silica (250 μm thick), which is a pure transparent quartz and acts as an insulator. This method maximizes the electric field through the use of new insulators in the accelerator structures. However, with normal insulators, the maximum electric field strength is limited by the formation of sparks, which arise because the electrons repeatedly bombard the surface creating an avalanche of electrons. Thus, to obtain a strong field accelerator, the formation of sparks must be presented by shortening the time during which the field is present. To decrease the insulation, the conventional high-voltage pulse of 1,000 ns to 1 ns leads to an increase in the surface cracking field of 5–20 MV/m. A new dielectric insulating configuration has enabled this limit to be increased to 100 MV/m. This high-gradient insulator is constructed using a row of floating conductors sandwiched between sheets of insulators. Thus, a DWA can be made by forming rings of HGI-material and additional conductive sheets at frequent intervals along the stack. Each of these blades is connected to a high-voltage circuit with a switch. When these switches are closed, an electric field is produced in the inner side of the HGI ring. By successive closures of the switches along the stack, the region of the strong electric field is changed along the stack, and protons traveling in phase with the wave will be accelerated through these rings. This arrangement accelerates protons to 200 MeV in a system that is 2 m in diameter (Fig. 3.12).

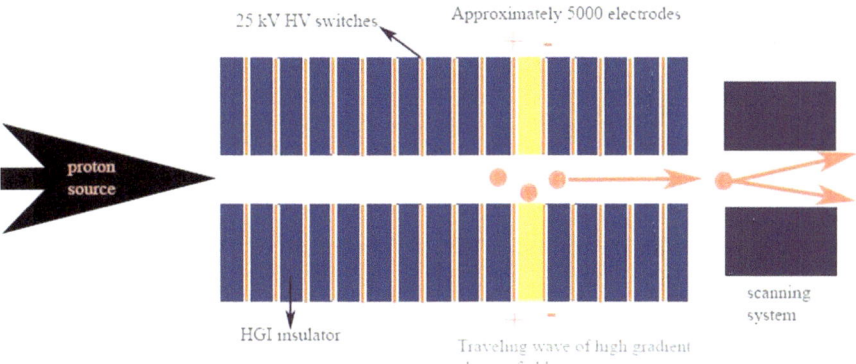

Fig. 3.12 Operating principle of the dielectric wall accelerator

Use of the DWA would circumvent some of the problems associated with conventional accelerators such as their expense and enormous size; the DWA costs 20 million US dollars and is much smaller than the accelerators used in the medical field. However, the DWA currently requires several improvements because of the high energies involved; these include improvement of the high-gradient insulators. As compared with other proton accelerators, the DWA is apparently the only accelerator for which the power, intensity and beam spot size can be varied pulse by pulse. The cyclotron only allows variation of the intensity under these conditions and the synchrotron allows variation of the energy and intensity, but not the spot size. The DWA allows variation of all these factors, pulse by pulse. Another advantage of this small linac is that it would be possible to mount it on a tomotherapy system. The Compact Particle Acceleration Corporation (CPAC) is developing a compact system for proton therapy based on the DWA that is much smaller than conventional accelerators, and is more powerful and very flexible. The idea of a compact proton accelerator comes from a team led by George Caporaso of the Livermore Beam Research Program at the Physical and Life Sciences Directorate. Their HGI is built with layers made of metal such as stainless steel alternating with layers of insulating plastic, such as polystyrene.

An induction accelerator formed by a set of HGIs can maintain extreme voltages. A particle injector starts the action, and the transmission lines made of dielectric materials and embedded conductors produce the electric field that drives the particles along the tube. The transmission lines are called Blumleins (named after British inventor Alan Blumlein). A laser supplies power to switches in the Blumleins through a distribution system consisting of optical fibers. The small solid state silicon carbide optical switches open and close at high speeds to control the high voltage that reaches each Blumlein, increasing the energy of the particles as they traverse the tube. The opening and closing of each switch creates a virtual traveling wave that pushes the energized particles along the tube (Fig. 3.13).

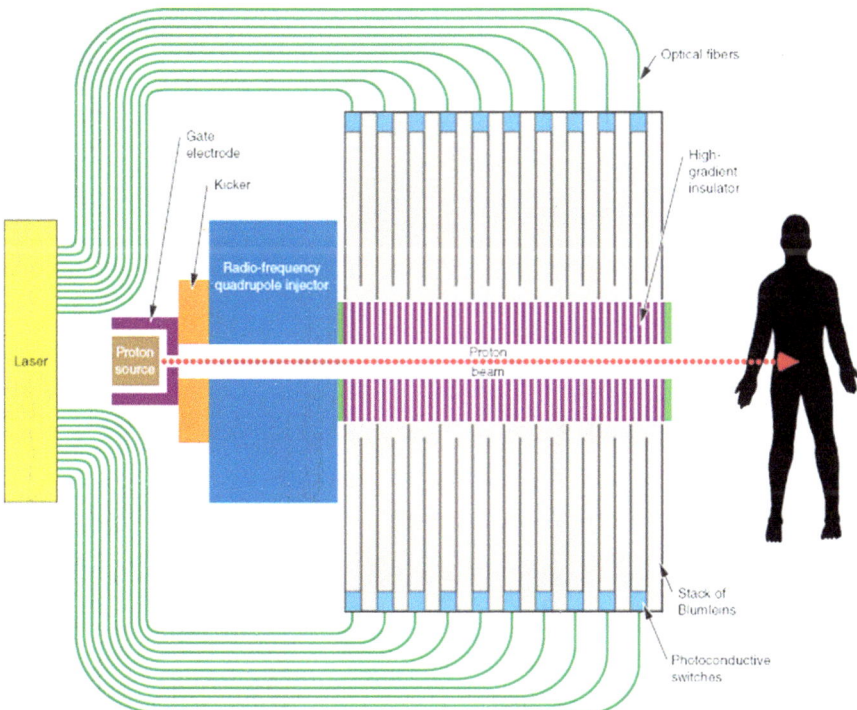

Fig. 3.13 Diagrammatic representation of the flow of protons in the dielectric wall accelerator from their source to the patient. Protons are sent to the interior of a "kicker" that injects them in pulses into a radio-frequency quadrupole, which compresses them into small bunches. Switches along the accelerator open and close at high speeds to control the voltage and increase the energy of the particles. Careful control of the switch mechanism creates a beam pulse with the velocity, shape, amplitude and length required for a given patient

This advance in size and power is due to three inventions: (1) the high-gradient insulator which allows a substantial increase in voltage-holding capacity; (2) the optical switches which can handle high-power loads at high speeds in a very compact size; and (3) the dielectric materials with embedded nanoparticles which facilitate the transmission and isolation of extremely high voltages (Fig. 3.14).

Fig. 3.14 George J. Caporaso examines the Compact Particle Acceleration Corporation's (CPAC's) newest Blumlein design. Tests at the CPAC are combined with computer simulations at Livermore Laboratory to produce a practical design. Courtesy George J. Caporaso

Thus, the CPAC estimates that it will be able to create a system for proton therapy that is accessible to all cancer treatment centers and their patients. The DWA can be used to accelerate electrons, protons or any ion, but more time is required before a clinical system can be established. Figure 3.15 shows a diagrammatic representation of the DWA.

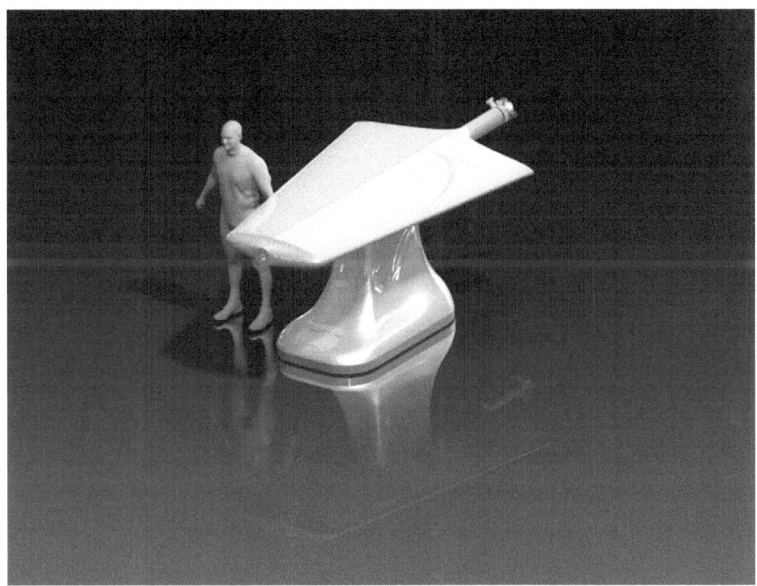

Fig. 3.15 An artist's rendition of the DWA in its fully developed form. Courtesy Dr. Anthony Zografos, Chief Operating Officer, Compact Particle Acceleration Corporation

3.10 Progress in Radiotherapy

The various types of radiation therapy modalities that have been used clinically or that are now under development are detailed in Table 3.1.

Table 3.1 Established radiation therapy modalities and new modalities under development

Year	Process	Energy (MeV)
X-ray radiotherapy		
1905	X-ray tube	0.05–0.45
1947	Van de Graaf	3
1948	Betatron	25
1953	Linear accelerator	4–25
Gamma-ray radiotherapy		
1910	Radium needles	1–3
1951	Cobalt "bomb"	1.17, 1.33
Electron radiotherapy		
1947	Van de Graaf	3
1948	Betatron	25
1961	Linear accelerator	4–25

(continued)

Table 3.1 (continued)

Year	Process	Energy (MeV)
Neutron therapy		
1969	Cyclotron	30
1975	Deuterium-tritium accelerators	14
1968	Californium-252	–
Proton therapy		
1955	Cyclotrons, Synchrotrons, Linear accelerators	60–250
New therapies under development		
Boron neutron capture radiotherapy (BNCT)	Nuclear reactors, Linear accelerators	–
Light ion radiotherapy	Cyclotron, Synchrotron, Cyclinac	400
Laser	–	–
Dielectric wall accelerator	–	–

3.11   Radiation Dose

3.11.1   Absorbed Dose

The absorbed dose is the energy deposited per unit mass in the medium (target) during exposure to ionizing radiation. It is measured in Gray (Gy), and 1 Gy = 1 J/kg. Thus, 1 Gray represents the absorption of 1 J of radiation per 1 kg of matter using the International System of Units. Because 1 J = 10^7 ergs and 1 kg = 1,000 g, 1 Gy = $10^7 \times 10^{-3}$ erg/g = 10^4 erg/g = 100 cGy.

3.11.2   Equivalent Dose

The equivalent dose takes into account the different biological effectiveness of different types of radiation. It is calculated by multiplying the absorbed dose by a radiation weighting factor appropriate to the type and energy of the radiation and can be calculated using the following equation:

$$H_T = \Sigma\, W_R\, D_{T,R},$$

where H_T is the equivalent dose absorbed by tissue T, $D_{T,R}$ is the absorbed dose in tissue T from radiation type R, and W_R is the radiation weighting factor. H is measured in Sievert (Sv) using the International System of Units, named after the Swedish physician Rolf Maximilian Sievert (1896–1966) who studied the biological effects of radiation. The weighting factor is estimated using the relative biological effectiveness (RBE) value of a given radiation type at low doses and not high doses. In the case of high doses the International Commission on Radiological Protections (ICRP) recommends a different definition of the equivalent dose:

$$H_e = \Sigma \, RBE_{Re} \cdot D_R,$$

where H_e is the equivalent dose and the weighting factors are replaced by the RBE value for a given radiation type (R) using a specific end point (e). The unit used here for dose is the Gray equivalent (GyE).

3.11.3   Effective Dose

The effective dose takes into consideration the dose due to ionizing radiation delivered non-uniformly. It takes into account both the type of radiation and the nature of each organ being irradiated. For a non-uniform radiation exposure, a different tissue weight factors (W_T) is used to reflect the different radiogenic sensitivity from different organs. The effective dose is given in Sv and is calculated as follows:

$$H_E = \Sigma \, W_{T.} H_T,$$

where H_E is the effective dose, W_T is the tissue weight factor defined by regulation, and H_T is the equivalent dose absorbed by tissue T.

3.11.4   RBE

When high doses are employed in particle therapy, the GyE unit has been used for heavy ions, while researchers working with proton therapy still use the cobalt Gray equivalent (CGE). The International Commission on Radiation Units and Measurements (ICRU) proposed replacing this unit with RBE-weighted absorbed dose, defined as $D_{RBE,V} = RBE.D_V$. The unit is given in Gy (RBE). The volume (v) must be specified and may correspond to the tumor volume or the planned target volume.

3.11.5   Integral Dose

The integral dose is defined as the average dose deposited in the total irradiated volume of the patient, when comparing different radiation therapy types. It is the product of the mass of tissue irradiated and the absorbed dose, and is given by the equation:

$$ID = m.D$$

where ID is the integral dose, m is the mass and D is the absorbed dose. The unit used is Kg × Gy.

3.11.6 Isoeffective Dose

The isoeffective dose D_{IsoE} is the dose of a radiation treatment carried out under reference conditions that produces the same clinical effects on the target volume as those of the actual treatment. It is the product of the total absorbed dose (in Gy) and a weighting factor W_{IsoE} defined by the equation $D_{IsoE} = D \times W_{IsoE}$. The weighting factor takes into account all clinical factors that could influence the clinical effects: the dose per fraction, the total time, the quality of radiation, the biological system and effects. In fractionated conventional X-ray radiation therapy, the dose per fraction and the overall treatment time are the two main parameters that can be adjusted. The weighting factor for an alteration of the dose per fraction is commonly evaluated using the linear-quadratic (α/β) model. For therapy with protons and heavier ions, radiation quality has to be taken into account. The isoeffective dose for heavy ion therapy is problematic because of the complex RBE field [11].

3.12   Units and Constants

3.12.1   Energy

The unit of energy used throughout is the eV (electron volt) as well as the following related units: MeV (mega electron volt), 1 MeV = 10^6 V; the GeV (giga electron volt), 1 GeV = 1,000 MeV; and TeV (tera electron volt), 1 TeV = 1,000 GeV. One eV is defined as the energy that an electron gains as it passes through an electric field with a potential difference of 1 V. The electron volt is a unit of energy equal to 1 eV ($1,602 \times 10^{-19}$ J); 1 joule = 10^7 erg, which is equal to 1 watt-s (1/3,600,000 kWh). As examples, electrons impact the screens of our televisions with an energy of a few keV (kilo electron volts). The initial particle accelerators had energies of up to 16 MeV. Currently, the LHC at CERN works with beams of 7 TeV and the total energy of the two beams is 14 TeV. In addition, cosmic ray particles with energies of up to 10^9 TeV have been detected.

If the speed of light and the reduced form of Planck's constant are both given values equal to one and then inserted into the equations used by particle physicists, the calculations and interpretations will be considerably simplified. To be more specific, the speed of light in a vacuum is approximately 3×10^{10} cm/s (more precisely c = 299792.458 km/s, or $2,998 \times 10^8$ m/s). Because E = mc^2 (Einstein's famous equation), mass and energy are essentially equivalent, and in principle the energy unit can be used as a unit of mass. If c = 1 is used in this equation, the mass can be directly expressed in eV. For the mass of an elementary particle, we employ MeV or

GeV. Regarding Planck's constant (h), the value of this constant is $4{,}135{,}669 \times 10^{-21}$ MeV. It is known from quantum mechanics that $E = h\nu$. E is the energy of a photon with frequency ν. Thus, both light and radio waves of a specific frequency ν consist of photons that have an energy $E = h\nu$. This equation together with that of Einstein's allows the establishment of a scale for all quantum phenomena. Planck's constant divided by 2π is what is called the reduced form of Planck's constant ($1{,}055 \times 10^{-34}$ J $= 6{,}582 \times 10^{-22}$ MeV). Usually it is represented by a cut h. It is found everywhere in quantum mechanics. If the reduced form of Planck's constant is set to one, we need only select units of length and time, so that the unit becomes one. In summary, working with the speed of light and the reduced form of Planck's constant as a unit facilitates work in particle physics; these are called natural units. When working at the macroscopic level, natural units are not convenient.

3.12.2 Mass

$$\text{Electron} \rightarrow m_e = 9.109 \times 10^{-31} \text{ kg} = 0.511 \text{ MeV/c}^2$$
$$\text{Proton} \rightarrow m_p = 1.673 \times 10^{-27} \text{ kg} = 938.27 \text{ MeV/c}^2$$

3.12.3 Charge

$$\text{Electron } e = 1.602 \times 10^{-19} \text{ C}$$

Finally, when working with very small particles, it is necessary to employ the following notations.

Before the decimal point:

deca	hecto	kilo	mega	giga	tera	peta	exa	zetta	yotta
10	10^2	10^3	10^6	10^9	10^{12}	10^{15}	10^{18}	10^{21}	10^{24}

For negative powers with zeros inserted after the decimal point:

deci	centi	mili	micro	nano	pico	femto	atto	zepto	yocto
10^{-1}	10^{-2}	10^{-3}	10^{-6}	10^{-9}	10^{-12}	10^{-15}	10^{-18}	10^{-21}	10^{-24}

3.13 Perspectives

If the equipment used in hadron therapy was less expensive, it could certainly be used universally in the treatment of cancer. In addition to the equipment, it is necessary to consider the cost of the associated building and ground area required for hadron treatment. A center that uses hadron therapy usually has 4–5 rooms. Below are detailed expected costs, which do not include specialized personnel, technicians, maintenance costs, insurance and other costs:

1. A proton accelerator costs about 20 M euros.
2. An accelerator for carbon ions costs 40 M euros.
3. A gantry costs at least 10 M euros. The gantry used at the HIT (Heidelberg) weighs 670 tons and allows precise rotation of the beam around the patient, and is thus an extremely useful device.
4. The ground area required for installation is 3,000 m², which would probably cost 20–40 M euros.

In conclusion, the investment required may reach 150 M euros, which is a cost that has seriously hampered the widespread use of hadron therapy.

A way to reduce these costs would be to eliminate the gantries and use only single-room facilities. If the gantry is not used, the irradiation has to be performed using a horizontal, vertical or inclined beam. Companies that sell equipment for proton therapy offer a gantry to allow irradiation at any angle. Only horizontal irradiation is used in the treatment of tumors located in the eye, head or neck. For some areas of the body where there is significant movement, for example the lungs (caused by respiration), irradiation in multiple positions is required necessitating use of the gating technique. Vertical irradiation is also used. Robotic systems are being developed to maintain the patient in different positions.

Researchers are working hard to create less expensive systems for accelerating protons and carbon ions. The cyclinac has enabled the use of particle beams with high energies, including carbon ions. It has also made the system more effective when compared with the synchrotron. Development of the cyclinac has allowed studies with gating and lowered the exorbitant costs associated with the use of the synchrotron. However, the costs are still high and consequently new emerging technologies are needed. Therefore, the DWA was developed in a collaborative project involving the Lawrence Livermore National Laboratory and Proton International [3]. The DWA is an induction accelerator, whose operation has been described earlier in this chapter. Issues that are currently being resolved regarding this system include the focusing of the accelerated protons and the viability of using a gradient of 100 MV/m, which would make it possible to rotate the DWA around a patient in a small single-room facility.

Regarding the use of a laser, it is expected that in the near future it may be possible to use a power of 10^{18} to 10^{20} W/cm² and short laser pulses of 30–50 fs. Companies are reducing the size and cost of high-power lasers. There are also projects for improving the beam quality. Several more years are needed to achieve these goals.

According to Amaldi et al. [3], the psychological and economic problems for the patient and the health service can be minimized by using a carbon ion beam in 4–5 sessions ("boost"). This can be given 1 week before the start of conventional radiation therapy at the radiation therapy facility nearest the patient's home.

References

1. Amaldi U (2007) Hadrontherapy: applications of accelerator technologies to cancer treatment. TERA Foundation Conference Presentation (17.05.2007). Available from: http://basroc.rl.ac.uk/basroc_files/icpt/.../RAL-Amaldi-17.5.07.pdf
2. Amaldi U, Braccini S, Citterio A et al (2009) Cyclinacs: fast-cycling accelerators for hadrontherapy. arXiv: 0902.3533 (physics.med-ph)
3. Amaldi U, Bonomi R, Braccini S et al (2010) Accelerators for hadrontherapy: from Lawrence cyclotrons to linacs. Nucl Instrum Methods Phys Res A 620:563–577
4. Bulanov SV, Esirkepov TZ, Khoroshkov VS et al (2012) Oncological hadrontherapy with laser ion accelerators. Phys Lett A 299:240–247
5. Chao AW, Chou W (eds) (2009) Reviews of accelerator science and technology, vol 2, Medical applications of accelerators. World Scientific, Singapore
6. Fourkal E, Shahine B, Ding M et al (2002) Particle in cell simulation of laser-accelerated proton beams for radiation therapy. Med Phys 29:2788–2798
7. Giudice GF (2010) A zeptospace odyssey: a journey into the physics of the LHC. Oxford University Press, Oxford
8. Grözinger SO, Bert C, Haberer T et al (2008) Motion compensation with a scanned ion beam: a technical feasibility study. Radiat Oncol 3:34
9. Lincoln D (2009) The quantum frontier: the large hadron collider. The Johns Hopkins University Press, Baltimore
10. Murata K, Mitsuoka K, Hirai T et al (2000) Structural determinants of water permeation through aquaporin-1. Nature 407:599–605
11. Newhauser WD, Durante M (2011) Assessing the risk of second malignancies after modern radiotherapy. Nat Rev Cancer 11:438–448
12. Schippers JM, Lomax AJ (2011) Emerging technologies in proton therapy. Acta Oncol 50:838–850
13. Surdutovich E, Obolensky OI, Scifoni E et al (2012) Superconducting synchrotron project for hadron therapy. Springer, New York

Chapter 4
Simulations

4.1 Simulations in Hadron Therapy

Simulations are important in hadron therapy for tumors located in areas that do not allow invasive procedures, such as inoperable neck carcinoma (see Fig. 1.12), uveal melanoma and pediatric tumors. The simulations act as a precision instrument that can be employed in these cases [15, 17]. Simulations using the Monte Carlo method can take into account all of the physical effects caused by the interaction of the particle beam with body tissue. The Monte Carlo code Geant4 (geometry and tracking) is often used for application development. This code is widely employed at CERN [1, 2, 9, 10] to study interactions in high-energy physics. The simulations and the results obtained are compared with experimental data, to analyze the limitations of physical models. This approach is valuable in hadron therapy. One problem is the computational time needed to implement the Monte Carlo method [17]. A computer with high processing speed and a large memory is required, because depending on the type and complexity of the application, long processing times may be necessary. This feature is available in Brazil through access to the computers at the CESUP (the National Center for Supercomputing at the Federal University of Rio Grande do Sul [UFRGS]). Internationally, several universities allow the use of their mainframe computers. A computer system for dosimetry in radiotherapy is available in Brazil, known as SISCODES, in which depth-dose profiles and isodose curves can be generated and superimposed. This system was developed by Bruno Trindade at the Nuclear Engineering Department, Federal University of Minas Gerais (NRI/UFMG). Simulations of hadron therapy also use data from the Visible Human Project, which may also demand long processing times. Working with simulations requires high-quality hardware.

4.2   Microdosimetry Measurements

A tissue-equivalent proportional counter (TEPC) is a plastic sphere with a wall thickness of 1.27 mm and an internal diameter of 12.7 mm, which is equivalent to a few microns of a tissue sphere (Figs. 4.1 and 4.2).

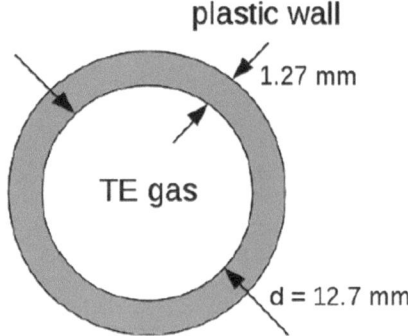

Fig. 4.1 Plastic sphere filled with gas at low pressure. *TE* tissue equivalent

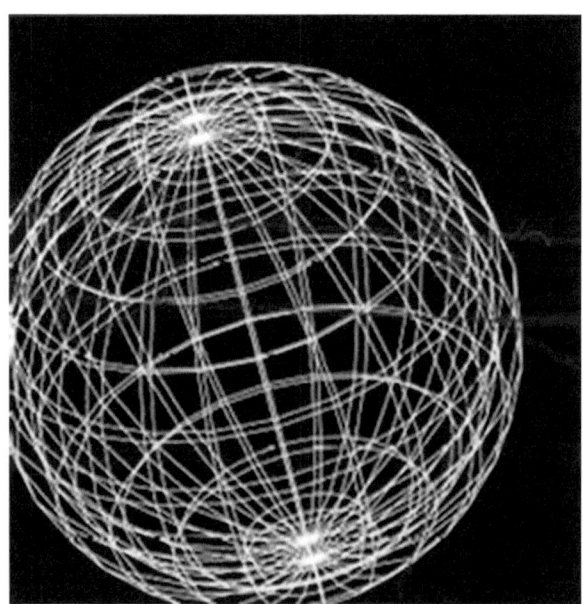

Fig. 4.2 Sphere with traces representing nucleons and nuclear fragments (*blue*). Fast electrons are shown in *red*

This device simulates a cell nucleus. The lineal energy is given by $y = \varepsilon/\lambda$, where y is the lineal energy, ε is the energy deposited in the TEPC and $\lambda = 2/3d$ is the average length. In a microdosimetry experiments, the lineal energy is measured using a TECP (Fig. 4.3).

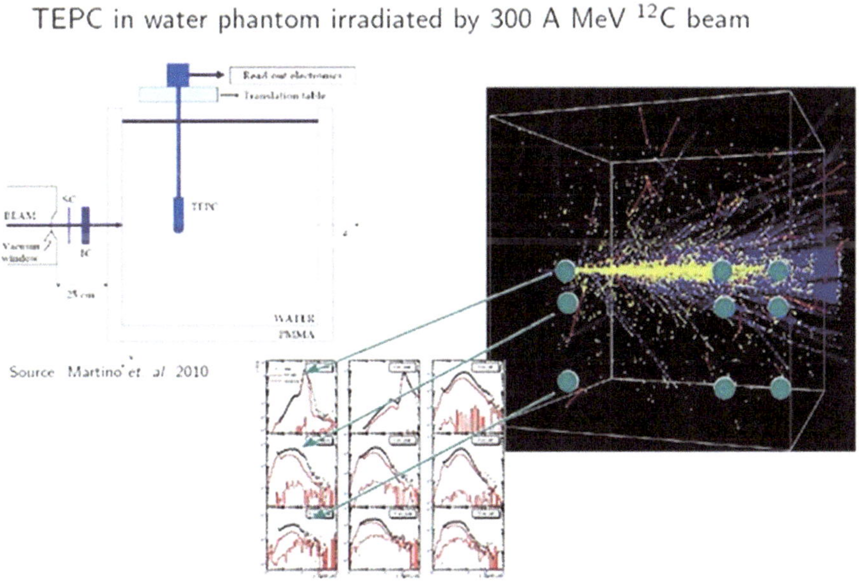

TEPC in water phantom irradiated by 300 A MeV ^{12}C beam

Fig. 4.3 Experimental arrangement used for microdosimetry at GSI [20]

The spheres are created to simulate cell nuclei, and have 1.27-mm-thick plastic walls and an inner radius of 12.7 mm. They are placed in water (phantom). The TEPC can be displaced on the translation table for any point XYZ in three dimensions (spatial). The measurements are performed by the electronics connected to the TEPC. To calculate the absorbed dose of radiation the following equation can be used:

$$D(Gy) = (0.204 / d^2) Y_{f,},$$

where d corresponds to the diameter of the simulated volume (μm), Y_f is given in keV/μm (average value), and D is the dose and is given in Gray (Gy). This unit is used in physical and not biological measurements.

4.3 Depth-Dose Profile

Figure 4.4 shows nine charts ordered according to the experimental arrangement for microdosimetry used at GSI Helmholtzzentrum für Schwerionenforschung in Germany.

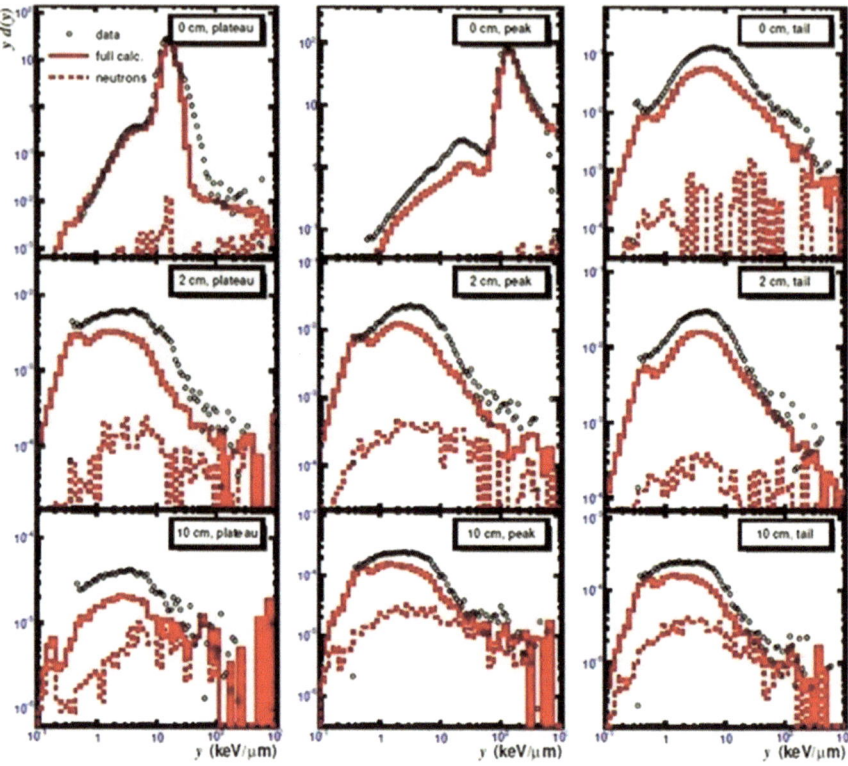

Fig. 4.4 Microdosimetry spectra simulation at various positions [20]

It can be seen in Fig. 4.4 that:

1. The contribution of neutrons increases when moving away from the beam.
2. The results obtained with the Monte Carlo model for Heavy Ion Therapy (MCHIT) agree well with the experimental data.
3. The spectra are systematically underestimated.

4.4   Fragmentation

Figure 4.5 shows the contributions of charged particles to the spectrum.

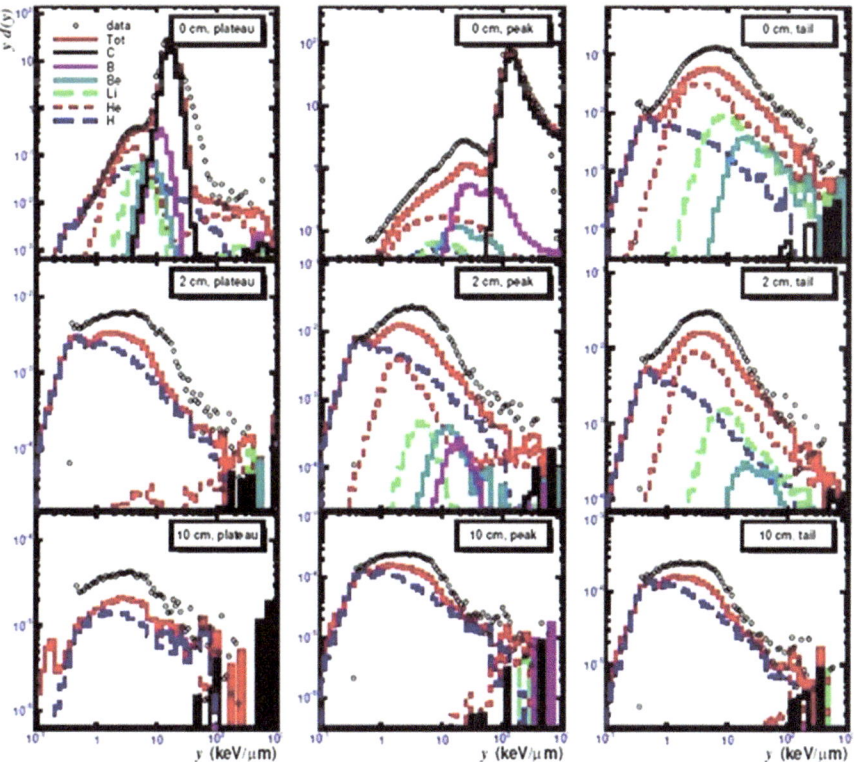

Fig. 4.5   Contributions of charged particles to the spectra [4, 20]

Analyzing the graphs, the following results can be arrived at:

1. The load and heavy fragments contribute, especially when they are near to the beam axis.
2. Spectra at 10 cm radius are caused by protons and neutrons.

Thus, important information can be obtained regarding cancer therapy using heavy ions, especially when coupled with Monte Carlo simulations. Monte Carlo codes can be used to:

- Calculate the dose distribution libraries at various levels of beam energy, as entries in planning systems for analytical treatment.
- Verify the results of the planning systems for analytical treatment, taking into account biological effects.
- Calculate doses outside the field.

The FLUKA MC code is used at the Heidelberg Ion Therapy Centre (HIT) for these purposes. Monte Carlo simulation is a slow process (h) relative to the planning system used for analytical treatments (min).

4.5 Calculations of Yields of Secondary Particles Using MCHIT

The search tool MCHIT was developed in the Frankfurt Institute for Advanced Studies (FIAS) in Germany to study ion transport through a medium similar to tissue, in the energy range used for cancer therapy with heavy ions (Fig. 4.6). The model is based on the application Geant4 software, which is used as a platform for the simulation of the passage of particles through matter. Geant4 is applied to the relevant experimental data in particle therapy. It works with phantoms and linear beam elements, but the same physical models can be used for Monte Carlo treatment planning.

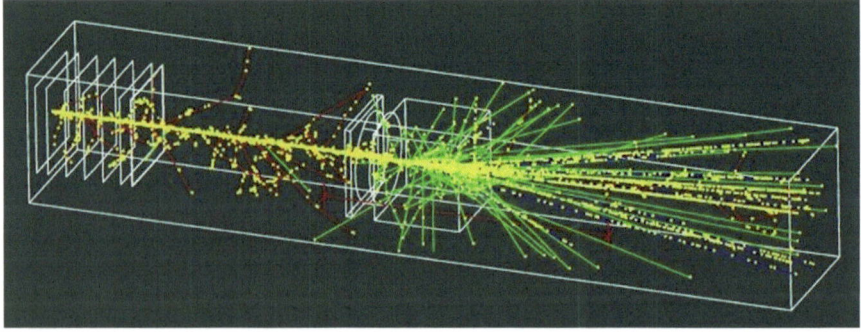

Fig. 4.6 Monte Carlo model for heavy ion therapy used at FIAS, Germany

In the physical processes of particle transport, the calculations take into account the following:

1. The loss of energy by ionization.
2. Multiple Coulomb scattering.
3. Nuclear fragmentation reactions.

The theory of "stopping power" was fully developed by 1933. The important fact to note is that protons stop in solid or liquid media. Beyond the stopping point, the radiation dose is negligible. The range of the proton is proportional to the square of its kinetic energy. If the proton beam is mono-energetic, all protons stop at approximately the same depth. The speed at which the proton loses energy increases when it slows down because for a given proton–electron collision, more momentum is transferred to the electron (the proton either remains in or leaves the neighborhood). Thus, the stopping power depends on the energy and material that slows it down. When corrected for density, material such as lead (Z=82) has less stopping power than materials such as beryllium (Z=4), water or plastics.

The term stopping power is commonly employed to signify the average energy loss per unit path length and is measured in MeV/cm. The stopping power depends on the type and energy of the particle and the properties of the material. The ionization density along a path is proportional to the stopping power of the material.

Accordingly, S(E)=−dE/dx, where S is the stopping power, E is energy and x is the path length. The minus sign makes S positive. The stopping power increases toward the end of the ions range and reaches a maximum, the Bragg peak, immediately before the energy drops to 0. The curve describing this is termed the Bragg curve. The deposited energy can be derived by integrating the stopping power over the entire length of the path of the ion.

Scattering theory was first published in 1947. In general, the deflection of a proton by a single atomic nucleus is weak and the angular scattering observed regarding a beam of protons is due to a random combination of many of these deflections. Because of this and the electromagnetic interaction, scattering is more properly known as multiple Coulomb scattering (MCS; Fig. 4.7). The spatial distribution is approximately Gaussian. The MCS theory accurately predicts the width of this Gaussian distribution because the energy of the proton, the type of material and thickness are known.

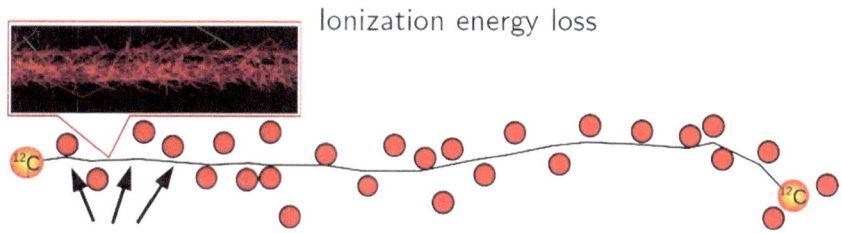

Fig. 4.7 Energy loss by ionization. Multiple Coulomb scattering (MCS)-multiple events

As mentioned, regarding the energy loss by ionization, −dE/dx corresponds to $1/\beta^2$ and equals 1/E (continuous drop: formulas, tables or approximations). With MCS formulas and approximations are used.

Nuclear fragmentation consists of simple events, and many secondary particles of several species are produced at large angles (this process is difficult to describe using a formula; Fig. 4.8).

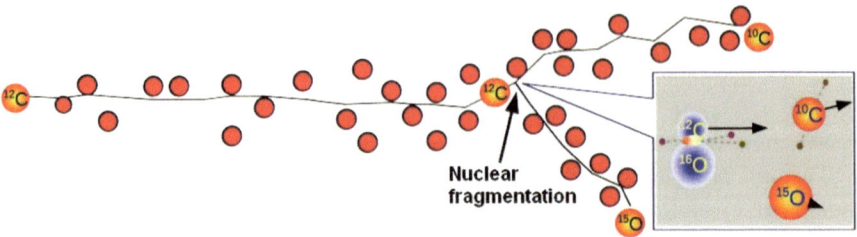

Fig. 4.8 Nuclear fragmentation

Below is shown a diagrammatic representation of a nucleus–nucleus collision (Fig. 4.9):

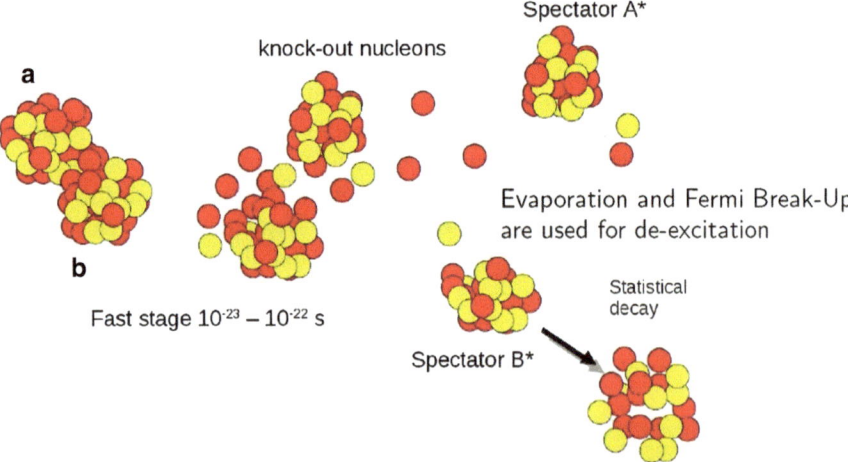

Fig. 4.9 Diagrammatic representation of a nucleus–nucleus collision

Now we consider the behavior of the depth-dose profile, with and without nuclear fragmentation. Figure 4.10 shows a marked overestimation of the dose at deep penetration without fragmentation reactions [21].

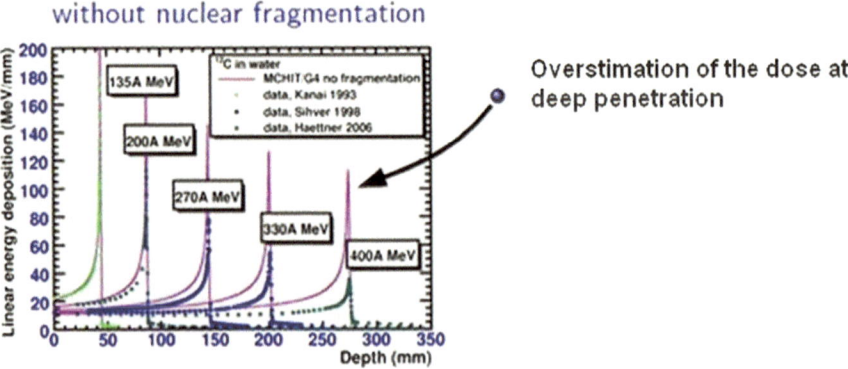

Fig. 4.10 Depth-dose profile without nuclear fragmentation [21]

When nuclear fragmentation is included, the peak height for deep penetration is reduced because of attenuation of the carbon beam. Extensions beyond the Bragg peak are caused by secondary fragments (Fig. 4.11).

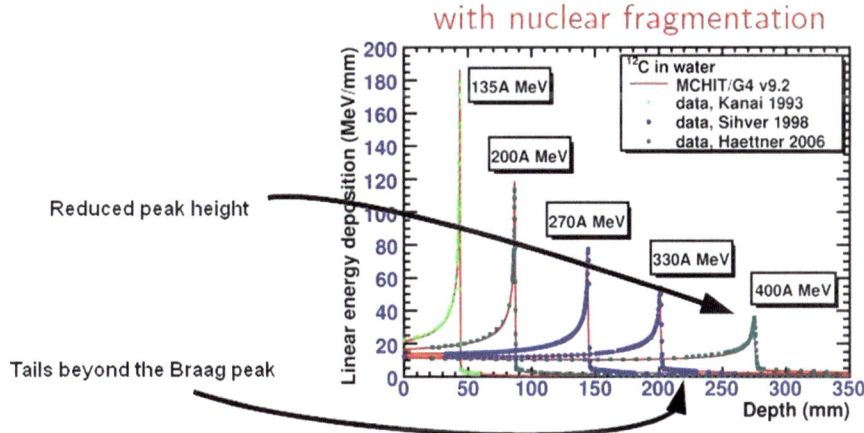

Fig. 4.11 Depth-dose profile with nuclear fragmentation [21]

Consider the fragmentation induced by a 400 MeV ^{12}C beam as shown in Fig. 4.12.

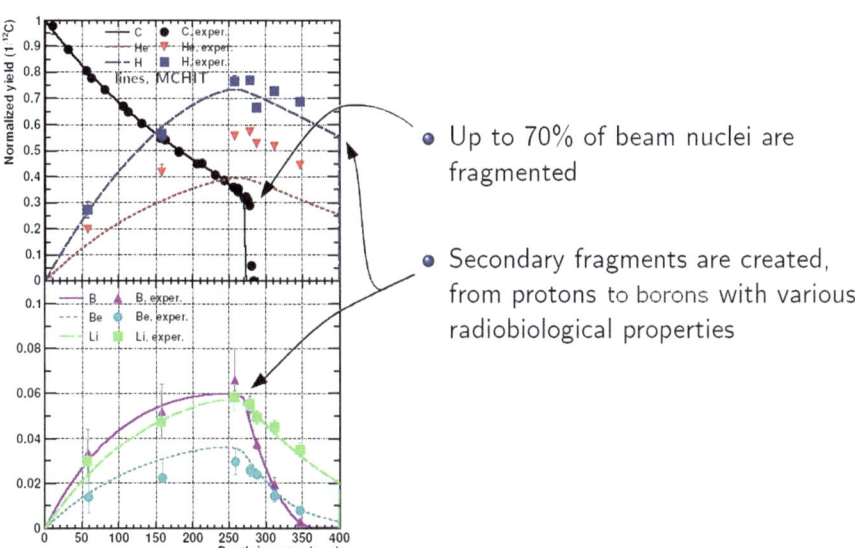

Fig. 4.12 Fragmentation caused by a 400 MeV ^{12}C beam

Up to 70 % of the nuclei from the beam are fragmented. The secondary fragments are created from protons to borons with various radiobiological properties. Much work is still required to develop models of nuclear fragmentation.

Figure 4.13 shows the production of secondary particles using MCHIT. One hundred events per 330 MeV ^{12}C beam occur in water (phantom) (Fig. 4.14).

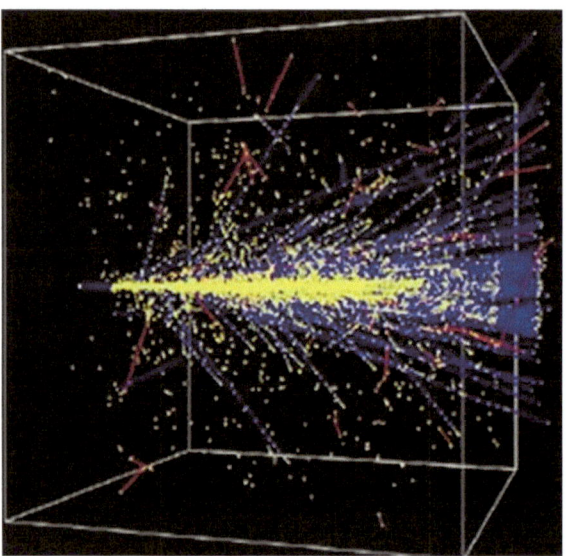

Fig. 4.13 Charged particles. *Blue*: nucleons and nuclear fragments; *red*: fast electrons

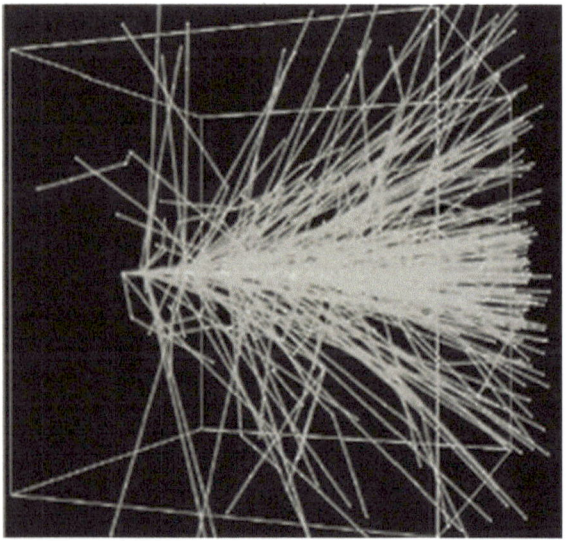

Fig. 4.14 Green: neutrons of all energies

4.6   Yield of Light Fragments: Angular and Energy Distributions

Figure 4.15 shows the production of light fragments with an angular distribution.

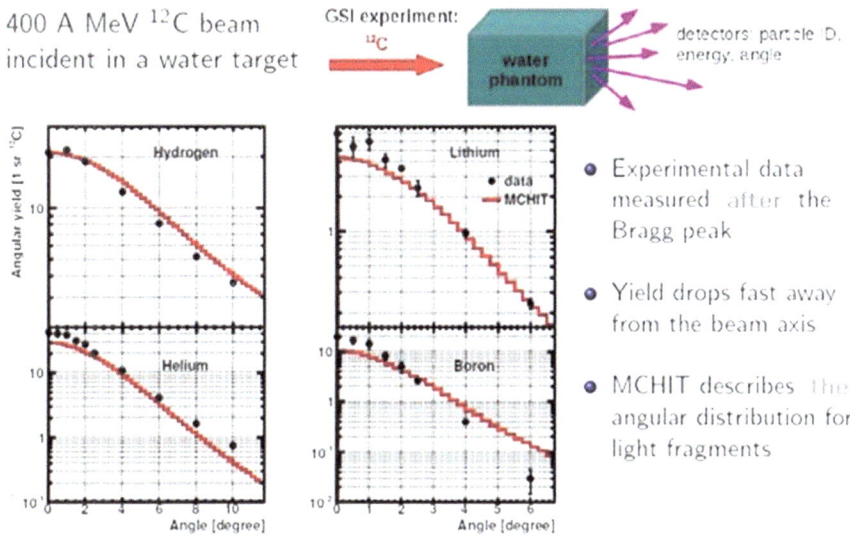

Fig. 4.15   Production of light fragments with an angular distribution [11]

Next the production of light fragments as a function of energy distribution will be considered (Fig. 4.16).

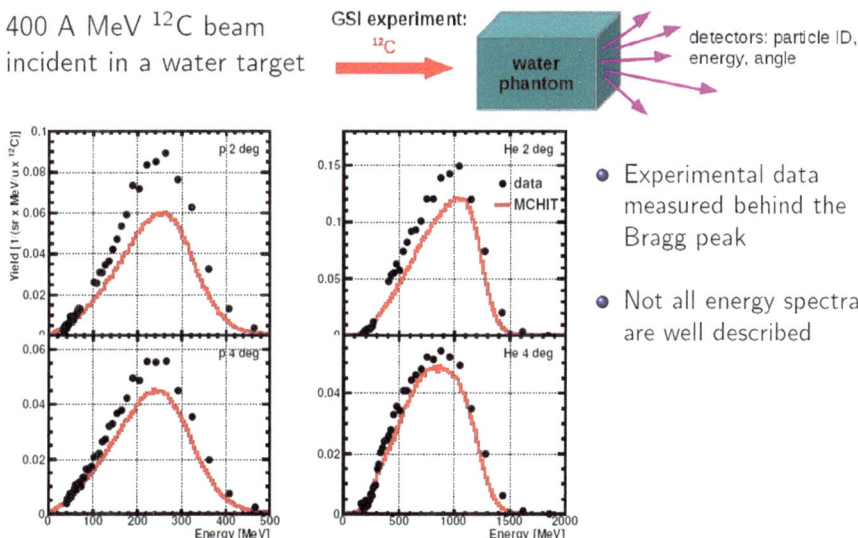

Fig. 4.16   Light fragments: energy distribution [11]

4.7   Impact of Radiation at the Microscopic Level

Jakob et al. [19] provided the image shown in Fig. 4.17, in which the proteins 53BP1 and RPA, both related to DNA repair, can be seen; they have been highlighted using fluorescent staining for better visibility. The authors employed a 9.5 A MeV ^{12}C beam to irradiate the specimen (a cell monolayer), which was visualized using a microscope.

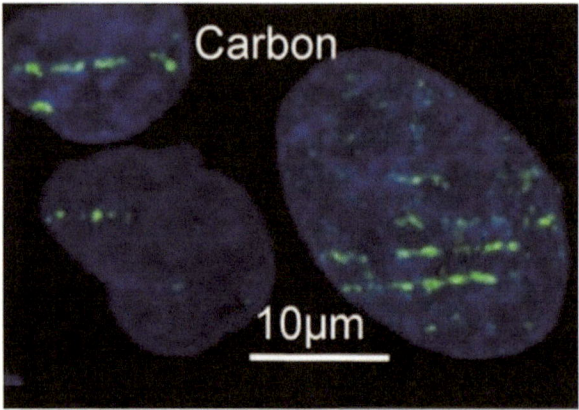

Fig. 4.17 DNA repair after irradiation. Proteins 53BP1 and RPA can be seen. They have been made fluorescent by immunostaining

The energy transmitted to the tissue per unit of track length is an important characteristic of radiation.

4.7.1   Conclusions

- The methods that are used to model detectors in experiments regarding nuclear and particle physics are also successful in simulating particle therapy (hadron therapy).
- The MCHIT model describes the transport of carbon nuclei with therapeutic energy into water (water phantom).
- The model describes the production of secondary nuclei from the fragmentation caused by the carbon beam.
- In general, microdosimetry spectra for ^{12}C beams are described by MCHIT.
- With this model the contribution to the dose from protons and neutrons outside the treatment field can be estimated.
- The MCHIT model can be extended for dose calculations around the ion track, on a scale of a few nm.

4.8   FLUKA: A Simulation Code

Internet links have been included in the references section of this chapter that provide open access to, and the use of, the simulation code FLUKA [8, 13–16, 18] to facilitate the work of readers interested in researching this code.

FLUKA was originally designed for physics research involving accelerators and detectors. Physicists use it to precisely predict electromagnetic and nuclear interactions in matter. For example, at CERN it is used to study beam-machine interactions and radiation damage. NASA has used it to analyze the radiation exposure of astronauts [3]. FLUKA is now used in state-of-the art therapy involving ion beam facilities, such as the HIT in Germany, to support treatment planning for cancer patients undergoing radiation therapy. Thanks to accurate models, FLUKA is employed to generate large amounts of data and provide access to commercial software used for treatment planning. It is also used for recalculating and verifying treatment plans. Till Böhlen, a researcher currently working at the CERN Partner Project is developing FLUKA for ion beam therapy. Böhlen states that FLUKA is a valuable tool to accurately compute treatment doses. This is particularly useful in critical treatment care situations, for instance if the patient has a metal implant in the target area of intervention. Future developments will include the development of improved FLUKA physical models for new ions, such as oxygen and helium, with a view to possible use in hadron therapy. The code is also widely used to simulate secondary radiation that is produced during treatment when patient tissues interact with the beam. This is essential because secondary radiation is being studied as a very powerful tool to perform in vivo monitoring during treatment (Fig. 4.18).

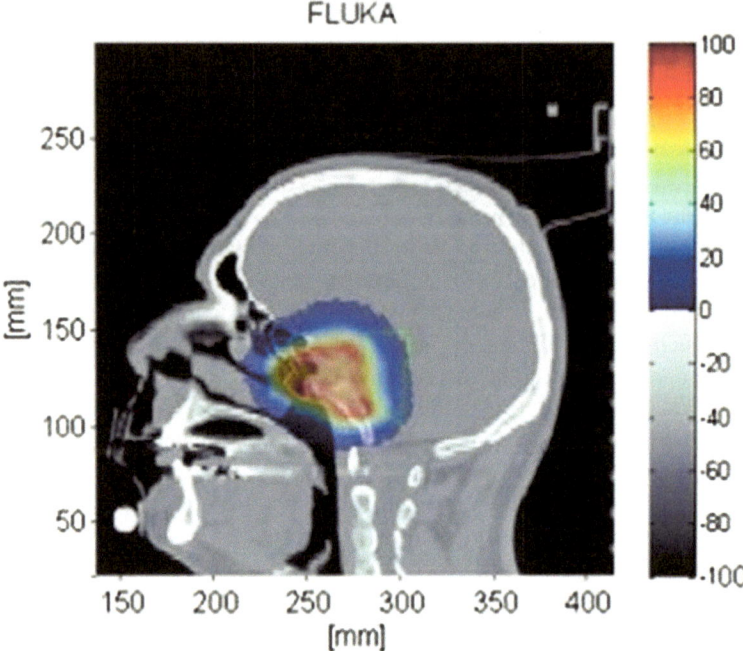

Fig. 4.18 Use of FLUKA to calculate the radiation dose distribution in a patient. The *color bar* shows the normalized values of the dose. Courtesy Andrea Mairani (CNAO, Italy)

In addition to FLUKA and MCHIT, GATE and GEANT4 software should be mentioned [1, 2]. GEANT4 has been used in applications involving particle physics, nuclear physics, accelerator design, space engineering and medical physics. It was created by physicists and software engineers using object-oriented technology and is implemented in the C++ programming language. GEANT4 is a code used to simulate the passage of particles through matter, encompassing geometry, physical models and tips, which are very useful in electromagnetic, optical and hadronic processes. It covers an energy range from 250 eV to TeV. GEANT4 has been widely used in simulations for hadron therapy. A literature search to check the comparative effectiveness with FLUKA highlights that it is still difficult to give any recommendation. Currently, it is proposed to first study the problem in more detail [5–7]. CERN comprehensively supports the development and use of FLUKA. It is used at the Centro Nazionale di Adroterapia Oncologica (CNAO), Pavia, Italy, where CERN researchers, mainly ENTERVISION fellows, seek experimental data for their theses.

References

1. Agostinelli S, Allison J, Amako K et al (2003) Geant4—a simulation toolkit. Nucl Instrum Methods A 506:250–303
2. Allison J, Amako K, Apostolakis J et al (2006) Geant4 developments and applications. IEEE Trans Nucl Sci 53:270–278
3. Andersen V, Ballarini F, Battistoni G et al (2004) The FLUKA code for space applications: recent developments. Adv Space Res 34:1302–1310
4. Ashton Acton Q (2012) Electrolytes: advances in research and application. Scholarly editions, Atlanta, Georgia
5. Battistoni G, Muraro, S, Salaet PR et al (2007) The FLUKA code: description and benchmarking. In: Proceedings of the Hadronic Shower Simulation Workshop 2006: AIP Conference Proceedings, vol 896, pp 31–49
6. Battistoni G, Cerutti F, Engel R et al (2006) Recent developments in the FLUKA nuclear reaction models. In: Proceedings of 11th international conference on nuclear reaction mechanism, Varenna, Italy, 12–16 June, pp 483–495
7. Böhlen TT, Cerutti F, Dosanjh M et al (2010) Benchmarking nuclear models of FLUKA and GEANT4 for carbon ion therapy. Phys Med Biol 55:5833–5847
8. Cerutti F, Battistoni G, Capezzali G et al (2006) Low energy nucleus–nucleus reactions: the BME approach and its interface with FLUKA. In: Proceedings of 11th international conference on nuclear reaction mechanism, Varenna, Italy, 12–16 June, pp 507–511. Available from: http://www.mi.infn.it/egadioli/Varenna2006/Proceedings/CeruttiF.pdf
9. Dementyev AV, Sobolevsky NM (1999) SHIELD—universal Monte Carlo hadron transport code: scope and applications. Radiat Meas 30:553–557
10. Giudice GF (2010) A Zeptospace Odyssey: a journey into the physics of the LHC. Oxford University Press, Oxford
11. Haettner E (2006) Experimental study on carbon ion fragmentation in water using GSI therapy beams. Master's Thesis, Kungliga Tekniska Hogskolan, Stockholm. Data available at EXFOR: http://www-ds.iaea.org/exfor/exfor.htm
12. http://www.fluka.org/fluka.php?id=licence
13. http://sourceforge.net/projects/auflukatools/
14. http://www.fluka.org/content/manuals/FM.pdf
15. http://www.fluka.org/fluka.php?id=license
16. http://www.isgtw.org/spotlight/physics-software-used-fight-cancer
17. http://www.lume.ufrgs.br/handle/10183/28650
18. https://www.fluka.org/fluka.php?id=secured_intro
19. Jakob B, Splinter J, Taucher-Scholz G (2009) Positional stability of damaged chromatin domains along radiation tracks in mammalian cells. Radiat Res 171:405–418
20. Martino G, Durante M, Schardt D (2010) Microdosimetry measurements characterizing the radiation fields of 300 MeV/u ^{12}C and 185 MeV/u ^{7}Li pencil beams stopping in water. Phys Med Biol 55:3441–3449
21. Mishustin I, Pshenichnov I, Greiner W (2010) Modelling heavy-ion energy deposition in extended media. Eur Phys J D 60:109–114

Chapter 5
The Next Challenge: Analysis of Motion in Radiotherapy

It is beyond the scope of this book to offer a complete review on the subject of motion in radiotherapy, but we can provide an overview of the fundamental concepts regarding techniques in use. Treating tumors located in organs involves patient motion, and inter- and intra-fractional movement of the organ itself. These movements affect each other. Scattering or scanning beam treatment modes can be used with particle beams. In the case of scattering, the concepts and data are often based on clinical use, that is, procedures that are known and have been used in practice for many years. In the field of scanning, the research is based on the effect of interference countermeasures. Currently, offsetting the dosimetric influence of target movement, both inter- and intra-fractional, is still an area of research relevant to hadron therapy. In general, inter-fractional movement occurs on a scale of minutes to hours, while in general, intra-fractional movement occurs on a scale of seconds to minutes. Examples of targets that exhibit inter- and intra-fractional movement are lung, liver and prostate. In charged particle beam therapy, the established methods that are used in photon beam therapy are employed to address these movements. Whatever the technique used to deliver the radiation, the measurement of depth-dose is strongly influenced by target movement. The published data have covered the adverse effects regarding movement of the target, because they can lead to under-dosing of the clinical target volume, even when field margins are used. For prostate treatment, intra-fractional movement is suppressed by means of dedicated immobilization, preventing the filling of the rectum and bladder by controlling drinking, or with enemas or a rectal balloon. The prostate moves more with a filled rectum than with an empty rectum. Inserting a balloon into the rectum, followed by inflation with saline solution, minimizes the amount of air in the rectum and distends its rear wall. The prostate also exhibits intrafractional motion [7]. Mayahara et al. [8], from the Hyogo Ion Beam Medical Center (HIBMC), instruct their patients to empty the bladder and rectum and do not use a rectal balloon for proton therapy.

The Rinecker Proton Therapy Center (RPTC) in Munich uses apneic oxygenation to treat tumors located in the thorax and high abdomen. Immobilizing the organ is difficult and often requires immobilizing the patient as a whole. To reduce respiratory motion, abdominal pressure is applied with a small plate, which causes a reduction in the movement of the diaphragm [11]. This system is used in the Heidelberg Ion Therapy Center (HIT) to treat hepato-cellular cancer using an ion beam scanner [10]. Reducing the breathing movements can be accomplished with special breathing maneuvers or artificial ventilation, for instance by holding the breath via respiratory control [5, 14]. Another alternative is apnea, which is used in the RPTC, for treating patients with tumors located in the lung and liver. Intubation using apneic oxygenation provides oxygen that does not require exhalation, and can thus reduce the amplitude of motion. For patient positioning, fluoroscopy is employed to check the consistency of the state of motion. Fluoroscopy is also used to check the variation in the range of motion, so that the beam aperture size is large enough and precise positioning is possible to ensure that tumor movement is restricted to the irradiation field.

Several research groups are working on developing techniques that allow coverage of patient motion during radiation therapy, including therapy involving the use of a scanned beam, by means of CTV. These include techniques involving gating, rescanning and beam tracking.

Gating is a technique that was developed for the radiation therapy of tumors that experience movement because of respiration [9] (Fig. 5.1). Irradiation occurs only during specific portions of the breathing cycle because the accelerator and energy selection system support multiple on/off cycles. In cyclotrons, fast intensity modulation is supported by the control system's beam current. Dedicated extraction techniques have been used previously in synchrotrons to enable on/off cycles per beam pulse. Gating has been used predominantly for lung cancer and the treatment of hepato-cellular injury.

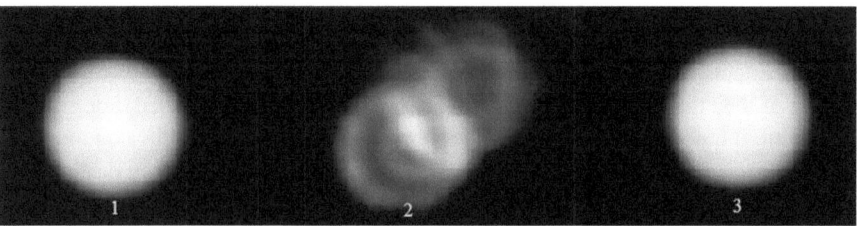

Fig. 5.1 (1) Fixed target, (2) moving target with no offset, (3) target with compensated motion (gating has been applied)

A system was developed at the Heavy Ion Medical Accelerator in Chiba (HIMAC) to control the beam during lung cancer treatment with the aim of minimizing the exposure of normal tissue in the vicinity of the tumor to radiation. The system for breath control during radiation therapy must meet the following conditions: (1) the signal for the commencement of irradiation needs to be precisely generated so that it only permits irradiation when the target is in the desired position; (2) the beginning and end of the beam extraction must respond quickly to a trigger signal; (3) the operation pattern of the ring requires optimization to provide the maximum effective radiation dose; (4) a shutdown system for the residual beam needs to be installed to prevent any undesired activation; and (5) provision of a security system for the treatment is necessary. To achieve this, the system employs a position-sensitive detector to detect movement of the target, an RF-knockout extraction with frequency and amplitude modulation, an optimized standard operating ring as used in the HIMAC synchrotron, a beam decelerator system to shut down the residual beam, and an interlock system for safe treatment [13]. Thus, gating increases the treatment time because the beam is frequently interrupted.

Rescanning is employed as a delivery option of the scanned beam in organs exhibiting intra-fractional motion. The principle is based on multiple irradiations at a proportionally reduced dose. A similar effect to rescanning could be achieved by the use of multiple fields for the delivery of treatment.

The beam tracking technique was proposed about 10 years ago by Keall et al. [6], and is currently being used to treat patients with the cyberknife synchrony system manufactured by Accuracy Inc. (Sunnyvale, CA, USA). Regarding ion beam therapy, Grozinger et al. [3, 4] proposed a beam-tracking system for a scanned carbon beam. They used a raster-scanner system for the lateral compensation of the components of organ movement and a linear motor to compensate for changes in radiological depth. This system was fully integrated into the GSI therapy control system by Bert et al. [1], Bert and Durante [2] and Saito et al. [12].

Treatment planning of tumor sites in motion involves the use of 3D computed tomography (CT). The dose is calculated using 4D CT in some institutions, but a lot of time is required for its routine use in the clinic; however, treatments with a scanned beam could benefit from calculating the necessary dose using 4D CT.

References

1. Bert C, Saito N, Schmidt A et al (2007) Target motion tracking with a scanned particle beam. Med Phys 34:4768–4771
2. Bert C, Durante M (2011) Motion in radiotherapy: particle therapy. Phys Med Biol 56:R113–R144
3. Grözinger SO, Li Q, Rietzel E et al (2004) 3D online compensation of target motion with scanned particle beam. Radiother Oncol 73:S77–S79
4. Grözinger SO, Bert C, Haberer T et al (2008) Motion compensation with a scanned ion beam: a technical feasibility study. Radiat Oncol 3:34
5. Hanley J, Debois MM, Mah D et al (1999) Deep inspiration breath-hold technique for lung tumors: the potential value of target immobilization and reduced lung density in dose escalation. Int J Radiat Oncol 45:603–611
6. Keall PJ, Kini VR, Vedam SS et al (2001) Motion adaptive x-ray therapy: a feasibility study Phys Med Biol 46:1–10
7. Langen KM, Willoughby TR, Meeks SL et al (2008) Observations on real-time prostate gland motion using electromagnetic tracking. Int J Radiat Oncol Biol Phys 71:1084–1090

8. Mayahara H, Murakami M, Kagawa K et al (2007) Acute morbidity of proton therapy for prostate cancer: the Hyogo Ion Beam Medical Center experience. Int J Radiat Oncol Biol Phys 69:434–443
9. Ohara K, Okumura T, Akisada M et al (1989) Irradiation synchronized with the respiration gate. Int J Radiat Oncol 17:853–857
10. Richter D, Steidl P, Trautmann J et al (2010) Mitigation of residual motion effects in scanned ion beam therapy. Radiother Oncol 96:S72
11. Rietzel E, Bert C (2010) Respiratory motion management in particle therapy. Med Phys 37:449–460
12. Saito N, Bert C, Chaudhri N et al (2009) Speed and accuracy of a beam tracking system for treatment of moving targets with scanned ion beams. Phys Med Biol 54:4849–4862
13. Wieszczycka W, Scharf W (2001) Proton radiotherapy accelerators. World Scientific, New Jersey
14. Wong JW, Sharpe MB, Jaffray DA et al (1999) The use of active breathing control (ABC) to reduce margin for breathing motion. Int J Radiat Oncol 44:911–919

ERRATUM TO

Hadron Therapy Physics and Simulations

Marcos d'Ávila Nunes

M. d'Ávila Nunes, *Hadron Therapy Physics and Simulations*, SpringerBriefs in Physics, DOI 10.1007/978-1-4614-8899-6, © Marcos d'Ávila Nunes 2014

DOI 10.1007/978-1-4614-8899-6_6

The publisher regrets that some figures and text sections were not fully cited in the original publication. The following figures are in whole or in part copyright of Lucas Burigo, and reproduced with permission:

Figure 4.1
Figure 4.2
Figure 4.3
Figure 4.4
Figure 4.5
Figure 4.6
Figure 4.7
Figure 4.8
Figure 4.9
Figure 4.10
Figure 4.11
Figure 4.12
Figure 4.13
Figure 4.14
Figure 4.15
Figure 4.16

The online version of the original book can be found at
http://dx.doi.org/10.1007/978-1-4614-8899-6

The following sections of text utilized the work of Lucas Burigo as a reference:

Section 4.4:
Thus, important information can be obtained regarding cancer therapy using heavy ions, especially when coupled with Monte Carlo simulations. Monte Carlo codes can be used to:

- Calculate the dose distribution libraries at various levels of beam energy, as entries in planning systems for analytical treatment.
- Verify the results of the planning systems for analytical treatment, taking into account biological effects.
- Calculate doses outside the field.

The FLUKA MC code is used at the Heidelberg Ion Therapy Centre (HIT) for these purposes. Monte Carlo simulation is a slow process (h) relative to the planning system used for analytical treatments (min).

Section 4.5:
The search tool MCHIT was developed in the Frankfurt Institute for Advanced Studies (FIAS) in Germany to study ion transport through a medium similar to tissue, in the energy range used for cancer therapy with heavy ions (Fig. 4.6). The model is based on the application Geant4 software, which is used as a platform for the simulation of the passage of particles through matter. Geant4 is applied to the relevant experimental data in particle therapy. It works with phantoms and linear beam elements, but the same physical models can be used for Monte Carlo treatment planning.

In the physical processes of particle transport, the calculations take into account the following:

1. The loss of energy by ionization.
2. Multiple Coulomb scattering.
3. Nuclear fragmentation reactions.

As mentioned, regarding the energy loss by ionization, $-dE/dx$ corresponds to $1/\beta 2$ and equals $1/E$ (continuous drop: formulas, tables or approximations). With MCS formulas and approximations are used. Nuclear fragmentation consists of simple events, and many secondary particles of several species are produced at large angles (this process is difficult to describe using a formula; Fig. 4.8).

Section 4.7.1
4.7.1 Conclusions

- The methods that are used to model detectors in experiments regarding nuclear and particle physics are also successful in simulating particle therapy (hadron therapy).
- The MCHIT model describes the transport of carbon nuclei with therapeutic energy into water (water phantom).

- The model describes the production of secondary nuclei from the fragmentation caused by the carbon beam.
- In general, microdosimetry spectra for 12C beams are described by MCHIT.
- With this model the contribution to the dose from protons and neutrons outside the treatment field can be estimated.
- The MCHIT model can be extended for dose calculations around the ion track, on a scale of a few nm.

We expand a number of entries from the Appendix for further accuracy and information:

What is the intensity-controlled raster scan method (ICRM)?

Answer: ICRM involves tumor detection in 3D so that the correct dose of carbon ions can be delivered. Increasing the energy of the carbon ions in the beam increases the depth of tissue penetration. By using horizontal and vertical magnetic fields, the ion beam can be displaced to the left or right. Thus, the tumor can be scanned in 3D, pixel by pixel.

What are hadrons?

Answer: in the narrow definition in particle physics these are strictly the subatomic particles made from quarks, anti-quarks and gluons (baryons and mesons) subject to the strong interaction. In medical physics jargon, the definition is broadened to include all highly interactive particles which are used in "hadron therapy", ranging from protons and neutrons to pions and heavy ions (α, C and Ne).

What is a phantom?

Answer: a specially designed artificial system which represents a reasonably accurate model of a biological system under investigation. It is used in medical imaging to evaluate the performance of an imaging device. The water phantom shown in Fig. 4.3 is a physical tissue model which mimics stochastic energy deposition in a cell nucleus.

What is MCHITand what is Geant4?

Answer: MCHIT stands for Monte Carlo model for Heavy-Ion Therapy. It is a software application based on the Geant4 open-source library of computational toolscreated by the international high-energy physics community. GEANT stands for GEometryANd Tracking.

Appendix

Questions and Answers Regarding Hadron Therapy, with Additional Information

1. **Why is treatment with conventional radiotherapy, even intensity-modulated radiation therapy (IMRT), inappropriate when a second malignancy arises?**

   Answer: Because it is recommended that carbon ion therapy is given as an alternative to conventional radiation therapy for such malignancies. Therefore, if a second course of conventional radiotherapy is undertaken, even using IMRT, it prevents other potentially more effective therapies from being administered; this could have serious consequences for the patient, including non-acceptance for treatment in hadron therapy centers in the USA. Furthermore, treatment using a carbon ion beam kills cancer cells that are resistant to X-ray or proton therapy and reduces the recurrence of cancer.

2. **Which are the best proton therapy centers in the world?**

   Answer: The M.D. Anderson Cancer Center's Proton Center, located in Houston, Texas is ranked as the best hospital for treating cancer in the USA (according to the US News and World Report, 2011). Loma Linda University Medical Center (California), Massachusetts General Hospital (Boston) and Wayne State University (Detroit) are also highly rated.

3. **Where is the best treatment center for carbon ion therapy located?**

   Answer: It is located in HIT, Heidelberg, Germany. The Japanese centers are also very good (HIMAC, HIBMC and GHMC). The CNAO in Italy recently received authorization from the Italian Ministry of Health to start operating. The center has synchrotron technology (PIMMS design developed by CERN) and active scanned proton and carbon ion beams; however, it lacks the financial resources to ensure their excellent operability. Researchers from CERN ENTERVISION use CNAO to obtain data for their research.

M. d'Ávila Nunes, *Hadron Therapy Physics and Simulations*, SpringerBriefs in Physics, DOI 10.1007/978-1-4614-8899-6, © Marcos d'Ávila Nunes 2014

4. **Why is there no hadron therapy in Brazil?**
Answer: This is because of several factors: (1) It costs up to 150 million euros to build an excellent synchrotron center. (2) The importance of this type of treatment, which has existed since 1943, has not yet been recognized. (3) It is cheaper to treat cancer patients with conventional radiotherapy.

5. **If a patient is not accepted at a given institution in the USA, should they look for another hospital?**
Answer: Yes, because standard protocols are different for each institution.

6. **What is PTCOG?**
Answer: This is the Particle Therapy Co-Operative Group. It organizes yearly scientific meetings and educational workshops regarding proton, light ion and heavy charged particle radiation therapy.

7. **How expensive is the equipment needed for hadron therapy? How much does a hadron therapy application cost? How many applications, on average, does a cancer patient require?**
Answer: The equipment for hadron therapy costs approximately US$ 80 million. Each application costs US$ 60,000. On average, 10 applications per patient are necessary.

8. **Will a health plan provider pay for hadron therapy? What is the current status in Brazil? What is the current status in the USA?**
Answer: In Brazil, to receive conventional radiotherapy, the patient has to file a claim. In the USA, all health plans pay automatically with no objections.

9. **Is there a center offering free hadron therapy?**
Answer: There are no centers worldwide that offer free hadron therapy.

10. **How do I contact LLUMC?**
Answer: To get in touch with LLUMC, contact the International Service of Loma Linda University Medical Center, c/o Christine Romero.

11. **How do I contact HIT?**
Answer: Before contacting HIT, it is necessary to fill out a specially developed form, containing all of the questions that the HIT doctors need answered before a treatment. This form can be found on the homepage of HIT. If the embassy of your country makes a request, there is no initial payment when you are hospitalized, but you will be billed after treatment.

12. **Should experts such as Professor Ugo Amaldi be contacted? What is the advantage in doing so?**
Answer: Yes. The advantage would be to gain the opinion and advice of an expert medical researcher in hadron therapy. The experience of the author in contacting these researchers has been highly beneficial.

13. **Where is hadron therapy offered?**
Answer: The centers listed in this book should be used, both inside and outside of the USA.

14. **What are the facts that are important to know concerning hadron therapy?**

Answer: A second course of conventional radiation therapy (even IMRT) should be avoided because this will prevent adequate treatment with carbon ions, with serious consequences. Hadron therapy is far superior to conventional radiotherapy.

15. **What resources are available in Brazil to assist patients in obtaining hadron therapy?**

Answer: There are no resources available in this regard in Brazil. Unfortunately, the hadron therapy centers in operation worldwide do not provide free treatment. The only recourse is to file a claim through a good lawyer who works in the healthcare sector.

16. **Does CERN offer scholarships for hadron therapy? What is a project partner? How does CERN participate in the dissemination of information regarding hadron therapy?**

Answer: Yes, CERN runs a PARTNER project that has major financial support to train doctors and researchers in hadron therapy, with more sophisticated features. CERN participates in the dissemination of information regarding hadron therapy through this project and also using the media. Applicants may also receive financial support from ENTERVISION.

17. **How is hadron therapy publicized in Brazil?**

Answer: Hadron therapy is publicized in Brazil through the internet, as we are doing, and through Masters' theses that are referenced on the internet. The pioneering attitude of Springer-Verlag in South America should be mentioned; they offered the opportunity to publish this Brief. President Dilma Rousseff has been petitioned to establish a proton synchrotron facility in Brazil. The request was forwarded to the Ministry of Health. We are joining efforts with the Barretos Cancer Hospital, the Cancer Hospital in São Paulo, and others to try to acquire such a facility; this will save countless lives and offer opportunities to work with researchers, across South America and worldwide, for the continued development of hadron therapy.

18. **Where was carbon ion therapy created?**

Answer: The original concept was developed at Berkeley in the USA and was subsequently taken up by Japanese researchers. Currently, there is an excellent carbon ion therapy center in Chiba, Japan. Carbon ion therapy is three times more efficient than proton therapy.

19. **What is the intensity-controlled raster scan method (ICRM)?**

Answer: ICRM involves tumor detection in 3D so that the correct dose of carbon ions can be delivered. Increasing the power of the carbon ion beam increases the depth of tissue penetration. By using horizontal and vertical magnetic fields, the ion beam can be displaced to the left or right. Thus, the tumor can be scanned in 3D, pixel by pixel.

20. **What is a cyclinac?**

Answer: A cyclinac is a combination of a cyclotron with a fast-cycling high-frequency linac, allowing for the generation of high energy (400 MeV) beams. This is useful for carbon ion therapy because these beams respond quickly to variations in energy (1 ms), which in turn facilitates work with respiratory gating. This system was developed at CERN by the TERA Foundation. The price of a cyclinac is lower than that of a synchrotron. A cyclinac is better than a synchrotron because the synchrotron responds slowly (1 s) to the power that causes variations in the system, and is therefore not suitable for studies involving respiratory gating. Because of its price, the cyclinac would be an excellent choice for proton and carbon ion therapy in Brazil, and would also help advance research in hadron therapy.

21. **Why is there no hadron therapy in South America? What facilities have been established in Argentina?**

Answer: For the reasons presented in the answer to question number 4, there is no hadron therapy in South America at present. In Argentina, a group is researching boron neutron capture therapy.

22. **What are the advantages of hadron therapy over conventional radiotherapy or even IMRT?**

Answer: Hadron therapy is far superior to conventional radiotherapy, even IMRT, because the beam energy is concentrated within the tumor and significantly reduces the possibility of tumor recurrence, in particular when using carbon ion therapy.

23. **What books and videos are recommended for the general public regarding hadron therapy?**

Answer: There are several videos on hadron therapy and particle accelerators available on the internet, as well as references. Some books are available through bookstores on the internet; most address proton therapy. Although carbon ion therapy is superior to proton therapy, in the USA only proton therapy is employed. It is interesting that the idea of using carbon ions for cancer treatment first originated in the USA, in Berkeley, California, but has subsequently been developed and used by Japanese scientists in Chiba, Japan. The USA embraced proton therapy, Japan carbon ion therapy. The Loma Linda University Medical Center uses a synchrotron to accelerate protons, but because the energy involved is only 200 MeV, it is not employed in carbon ion therapy.

24. **What are hadrons?**

Answer: Hadrons, are highly interactive particles such as protons, neutrons, pions and heavy ions (α, C and Ne).

25. **How are protons, neutrons and carbon ions obtained for accelerators (cyclotron and synchrotron)?**

Answer: (1) Protons are produced by applying an arc discharge in hydrogen gas using a source called a duoplasmatron. The electron is released from the hydrogen atom leaving the positive nucleus (a proton) floating freely in the resulting plasma. By applying an electric field, protons are extracted from the surface of the plasma and are sent on as a stream of positive particles. Thus, currents of up to 300 mA can be obtained. (2) Neutrons are obtained by accelerating deuterons with an energy of 48.5 MeV onto a beryllium target. The deuterons are accelerated using a cyclotron superconductor. Generally, neutrons can be obtained by accelerating protons (p) or deuterium (^{2}H) and colliding them with a beryllium or lithium target. This provokes reactions of the type ^{9}Be (p, n) ^{9}B, ^{7}Li (p, n) ^{7}Be, and ^{3}H (^{2}H, n) ^{4}He. (3) Heavy ions are atomic nuclei that are heavier than hydrogen nuclei that have lost their electrons. A variety of ions that are heavier than protons are used, among them helium nuclei, carbon and oxygen. Heavy ions are three times more effective than protons and helium ions. In the human body, heavy ions can be targeted with millimetric precision and are therefore superior to protons in the treatment of certain tumors. As is well known, ions are charged atoms. Thus, to obtain ions, the atoms must necessarily lose their negatively charged electrons. For this purpose, carbon dioxide gas flowing inside an ion chamber is used. Free electrons in the gas are accelerated by magnetic fields and microwaves. Traveling through the ion chamber, the electrons impact the molecules of carbon dioxide. After the collision, the molecules dissociate and four of the six electrons in the carbon atom are separated. Then, electric fields are employed to extract the carbon ions from the chamber. Special magnets transport them in a steady flow in a vacuum. This flow is converted into a regularly pulsating flow with a frequency of 217 million pulses per second. The beam is collimated and the ions are accelerated. Subsequently, electromagnetic fields accelerate the ions up to 10 % of the speed of light. Exiting the accelerator through a sheet of carbon, the carbon atoms lose their remaining two electrons, so that only the nuclei with six positive charges remain.

26. **How do charged particles penetrate the tumor? What is the Bragg peak? How do charged particle beams differ from conventional radiation therapy beams?**

Answer: Charged particles disrupt the strands of the DNA double helix in the nuclei of tumor cells; because of this damage, the tumor cells cannot replicate and die. The ionization density produced by a charged particle along its track increases as it slows down. It eventually reaches a maximum known as the Bragg peak close to the end of its trajectory. After that, the ionization density rapidly diminishes. Because of the Bragg peak the entire beam energy can be released within the boundaries of the tumor. Charged particles can be precisely confined within the tumor and do not significantly penetrate healthy tissue beyond the tumor. Hadron therapy produces a high rate of ionization in addition to damaging DNA. In contrast, conventional radiotherapy beams pass through normal tissue that lies in the path to the target tumor; they can cause unwelcome damage to healthy tissue.

27. **How do cyclotrons and synchrotrons work?**
    Answer: The cyclotron is a device invented in 1932 by Ernest O. Lawrence. It works with a magnetic field that imposes a circular path on the charged particles; these are subjected to an oscillating electric field with a fixed frequency that accelerates them. The final trajectory is a spiral. In the cyclotron, the time spent on the path is the same for all particles and is independent of radius. The synchrotron originated from the cyclotron; it differs in that it uses the principle of phase stability, maintaining the synchronism between the applied electric field and the revolution frequency of the particle. The magnetic field maintains the orbit of the particles rather than the process of acceleration; therefore, the magnetic field lines are only necessary in the annular region defined by the orbit.

28. **Can particles be accelerated using a laser?**
    Answer: New devices are being developed to accelerate particles using a laser. This could in principle reduce the costs involved.

29. **The first cyclotron had a diameter of only 15 cm. Why do the current cyclotrons weigh 220 tons and have diameters of several meters?**
    Answer: The current cyclotrons are so much bigger because a better stabilized magnetic field in the orbital path and higher power are needed to deliver higher energy to the beam. Current cyclotrons typically operate up to 250 MeV.

30. **Is there an economic advantage in transforming a synchrotron light source (such as the existing one in Campinas) to a proton synchrotron?**
    Answer: No, there is no economic advantage in transforming a synchrotron light source to a proton synchrotron. The expense entailed in replacing equipment is about the same as buying a new proton synchrotron. For more details please refer to the main text.

31. **How are particles accelerated using a laser?**
    Answer: It is possible to accelerate protons by means of a violent acceleration of electrons in the laser field, which then draws the protons onto the posterior surface of the target. A continuous spectrum of protons is obtained. Consider a powerful laser pulse acting violently on a target constituted by a thin blade doped with hydrogen atoms. The laser accelerates electrons off the target in its posterior region, creating an electric field favorable to the output of protons from the target. According to the literature, this emerging technology will soon be available for radiation therapy. The laser system is smaller, cheaper and easier to control than other systems. Laser systems generate beams of excellent quality. Several centers have already obtained beams of protons and other ions using a laser. A simple comparison between the traditional accelerators and laser accelerators indicates the relative viability of large lasers.

32. **Briefly, what was the history of hadron therapy and the pitfalls encountered in its development?**
    Answer: A brief history of hadron therapy is given in Chap. 1. The pitfalls encountered were many and severe. Even the Massachusetts General Hospital in Boston discontinued the clinical use of hadron therapy because the results were

not satisfactory; there were no significant improvements in therapeutic outcome using this treatment. However, research was continued in other centers. Currently, there are centers of excellence in proton therapy at LLUMC in California, and in hadron therapy with carbon ions at HIT, Germany and Chiba, Japan.

33. **What is the mechanism of action by which protons and carbon ions damage tumor cells? Why are ions that are larger than carbon ions not used in hadron therapy?**

Answer: Both protons and carbon ions penetrate tumor cells and disrupt the intracellular DNA strands, which prevents the cells from replicating and can also kill them. In general, ions are better for radio-resistant tumors, while protons minimize the risk of secondary tumors. Ions heavier than carbon ions should not be used because it has been shown that these can cause significant adverse effects in normal tissue.

34. **Why are hadron therapy simulations important? What is needed to carry out these simulations in terms of computers (hardware) and software?**

Answer: The simulations are required to verify which factors contribute to the deleterious effect of accelerated particles that penetrate the tumor cell. Simulations are common in various areas of research (Monte Carlo simulation, Geant4), because they identify responses to irradiation without the need to conduct the experiment itself. Very powerful computers are necessary for these simulations because of the amount of calculations required and for speeding up the operating time. The software is usually developed in C++, and often requires a large amount of computer memory and high processing speeds.

35. **How is the plateau that occurs along the length of the tumor in the depth-dose graph explained? For further application of the Bragg peak, what should be done (e.g. beam intensity increase)? Can the particle beam be moved to the right and left along the tumor (using electromagnets with the magnetic field oriented vertically or horizontally)?**

Answer: The plateau is the result of the summation of Bragg peaks obtained with deeper penetration of the particle beam within the tumor. The beam releases its energy after reaching the Bragg peak. Electromagnets with a horizontal or vertical magnetic field can shift the particle beam to the left or right. Thus, it is possible to sweep the entire volume of the tumor, pixel by pixel.

36. **In scientific studies of hadron therapy, terms such as Bragg peak, RBE, LET, SOBP and Gy are used. Can you explain what they mean?**

A: The Bragg peak has been explained previously. The RBE is the relative biological effectiveness, which is related to the spatial density of the energy deposition, known as linear energy transfer (LET). The physical dose delivered by photons (sparsely ionizing or low LET radiation) is more evenly distributed than an equal physical dose of heavy ions (dense ionization or high LET radiation). High-energy protons are similar to X-rays, but at low energy their LET increases and their RBE is higher than that of alpha particles, considering the same LET. The secondary low-energy protons are responsible for the high RBE of neutrons, which are candidates to be the most effective regarding the induction of particle late effects. The spread-out Bragg peak (SOBP) is the sum of all of the Bragg peaks, and represents the therapeutic radiation distribution within the tumor. Gray (Gy) is the unit of measurement of the absorbed dose (or simply dose) and is the unit of energy deposited per unit mass in the target. One Gy can also be defined as the amount of energy, expressed as joules, derived from the radiation that is absorbed per 1 kg of body weight.

Units and Measurements of Radiation

Below is a summary table showing the equivalent for each radiation unit, given as the old or special unit and the International System (SI) unit.

Magnitude	Old unity or special	Unity SI	Equivalent
Activity (A)	Ci (Curie)	Bq (Bequerel)	1 Ci = 37 G Bq
Exposition (X)	R (roentgen)	Gy (Gray)	1 Gy = 100 R
Radiation dose	Rem (roentgen equivalent in man)	Sv (Sievert)	1 Sv = 100 rem
Absorbed dose (D)	Rad (radiation-absorbed dose)	Gy (Gray)	1 Gy = 100 rad
Equivalent dose (H)		Sv (Sievert)	

37. **What is the unit of measurement for the radiation dose absorbed by humans? How is it measured? What is the maximum allowed?**

Answer: The unit for measuring the physical radiation dose absorbed by humans is the Gray (Gy). This is a unit in the International System of measurements. The Gray indicates how much ionizing radiation is absorbed in the target per unit mass; 1 Gy = 1 J/kg. The unit was named in honor of Louis Harold Gray, a British radiologist. It is considered that the same biological effect results from the absorption of 4 Gy gamma radiation or 0.2 Gy of alpha particles. These two different doses of radiation are said to be equivalent in their biological effects. The old unit of equivalent dose is the rem (roentgen equivalent in man). It is equivalent to a radiation dose whose effect is similar to the effect of 1 roentgen in man. 1 Gy of X-rays corresponds to about 100 rem. The dose tolerance for a nuclear-plant worker is 5 rem/year, and for residents in the vicinity of nuclear power plants it is 0.5 rem/year. A dose of 25–50 rem can cause minor blood changes and an increased risk of developing cancer. A whole body dose of 400 rem is the average lethal dose and will cause death in 50 % of the exposed population within 60 days. A massive whole body dose of 500 rem results in the death of up to 100 % of exposed individuals within 2 days. To summarize: 1 Gy = 1 J/kg. 1 Gy is deposited by 2×10^{10} protons if the protons are stopped within 1 kg of body tissue. Typically 1/2 to 2/3 of the energy is deposited outside of the tumor.

38. **Who invented the cyclotron, saved his mother from uterine cancer using radiotherapy and won the Nobel Prize for Physics in 1939 for his invention?**

Answer: The scientist was Ernest O. Lawrence.

39. **What causes the blue color in the central region of the Crab Nebula?**

Answer: The blue color occurs as a result of synchrotron type radiation, which is produced when high-energy particles are moving at high speeds, including electrons forced to travel in a curved path by a magnetic field. This radiation is proportional to the fourth power of the particle velocity and inversely proportional to the square of the radius of the trajectory.

40. **How is the cyclotron frequency calculated? How do you prove that the time spent by a particle in its path is the same for all orbits and independent of the radius?**

Answer: Inside the cyclotron, a single lap completed by a particle is equal to $2\pi r$ (trajectory radius r). If t is the time required to complete half a lap, then $v = \pi r/t$, and $t = \pi r/v$. Because $v = q.B.r/m$, then $t = \pi.m/q.B$. Thus, the time spent on the course is the same for all orbits and is independent of radius. Because the time period required for one complete lap is twice the time required for the half lap, $T = 2.t$, then $T = 2\pi m/qB$. The frequency (v is the inverse of the period), so $v = 1/T$, then $v = qB/2\pi m$.

Questions and Answers Regarding Simulations in Hadron Therapy

1. **What is the experimental setup used for microdosimetry? What is a water phantom?**

Answer: The experimental setup used for microdosimetry is extremely simple. A plastic ball filled with low-pressure gas is employed. This sphere has a wall with a thickness of 1.27 mm and an internal diameter of 12.7 mm. It simulates a cell nucleus. The spheres are placed inside a tank with water, and attached to a translation table that moves in directions X, Y and Z, or three-dimensionally. At the top of the table the translation has an electronic reading. A tissue equivalent proportional counter (TEPC) receives the radiation beam as outlined in Fig. 4.3.

The phantom is a device or system which represents a reasonably accurate model of a biological system under investigation. The Fig. 4.3 shows a phantom used to mimic cell nuclei under the action of radiation.

2. **How can proteins be observed repairing the DNA of cells undergoing radiation?**

Answer: Proteins can be observed repairing the DNA of cells undergoing radiation by employing a fluorescence staining technique for the 53BP1 and RPA proteins, which are both involved in DNA repair. Microscopy involving a beam of 9.5 A MeV ^{12}C is used.

3. **FLUKA is a simulation code. What other software can be used for the same purpose?**

   Answer: FLUKA is a simulation code used at CERN to study beam-machine interactions and radiation damage. NASA has used it to study the radiation exposure of astronauts. FLUKA is also used for simulations in hadron therapy, especially at HIT in Heidelberg, Germany, to support the treatment of patients with cancer undergoing radiation therapy. Other types of software used for the same purpose are GEANT4 and the MCHIT, developed at FIAS in Germany. The software is always developed using the Monte Carlo method.

4. **Why are simulations important in hadron therapy? Why are computers with high processing speeds and memory capacities required to perform these simulations?**

   Answer: Simulations are important to better understand the processes involved in hadron therapy and also in the calculation of radiation doses. The amount and complexity of the calculations required by the simulation codes are high, requiring the use of computers with large memory capacities.

5. **In the physical transport of particles, what should be taken into account in the calculations? Explain the factors involved.**

   Answer: The physical transport of particles takes into account the energy loss by ionization, multiple Coulomb scattering and nuclear fragmentation reactions. This is explained diagrammatically in Figs. 4.7 and 4.8.

6. **Explain from a molecular point of view why carbon ions are more effective than protons in the treatment of cancer. What are the effects regarding tumor recurrence?**

   Answer: Carbon ions are larger than protons and usually break the two strands that make up DNA, preventing its repair. The damaged cell does not replicate and dies. Other ions with different sizes have been tried, but carbon ions have been found to be the most effective. If a second course of conventional radiation is administered, and fails to prevent tumor recurrence, further treatment with radiation (including carbon ions) is not possible and the patient can only rely on chemotherapy. Treatment with carbon ions can prevent tumor recurrence (i.e., a second occurrence of malignancy) to varying degrees dependent on the tumor type and grade. Overall, carbon ions appear to be extremely effective in cancer therapy, but clinical data is relatively limited at this time. Figure 2.1 illustrates the breaking of one strand of DNA (top figure) and of two strands (irreparable).

7. **If a cancer patient has a metal implant in the target area of intervention, how is it possible to compute the radiation doses for treatment? Could the FLUKA software be useful in this case?**

   Answer: When a patient has a metal implant in the irradiation target area, doses for treatment are necessarily obtained using FLUKA software or analytical methods. FLUKA software is extremely useful in such cases.

8. **Why should Gray (Gy) be used as a unit for calculating the absorbed radiation dose?**

Answer: Gray is a physical unit rather than a biological one and is therefore the appropriate choice.

9. **When simulations are carried out involving microdosimetry spectra at various positions, does the contribution of neutrons increase or decrease moving away from the beam? Do the results obtained with MCHIT agree or disagree with the experimental data? Are the spectra systematically overestimated or underestimated?**

Answer: The contributions of neutrons increase when moving away from the beam because new impacts occur. The results obtained from the simulation agree well with the experimental data, but the spectra obtained using simulation are underestimated compared to experimentally obtained spectra.

10. **When analyzing the contributions to the spectra of charged particles, do charged and heavy fragments contribute more near the beam axis or away from it? Which particles cause the spectra at 10 cm radius?**

Answer: The contribution of charged and heavy fragments is greater near the beam axis. The spectra at 10 cm radius are mainly caused by protons and neutrons.